# Spring Boot+Spring Cloud+Vue +Element项目实战 手把手教你开发权限管理系统

徐丽健 著

清华大学出版社
北京

## 内 容 简 介

本书从项目实践出发,手把手、心贴心地带领读者从零开始,一步一步地开发出功能相对完整的权限管理系统,从而深入掌握当前主流的 Spring Boot + Spring Cloud + Vue 前后端集成开发技术。

全书分为三篇共 32 章。第一篇为系统介绍篇,对系统的功能、架构和界面进行介绍,对系统的安装运行给出指南,对涉及的关键技术进行简单介绍。第二篇为后端实现篇,从数据库设计和搭建开发环境开始,全面细致地讲解权限管理系统的后端实现全过程。第三篇为前端实现篇,从搭建开发环境开始,全面细致地讲解权限管理系统的前端实现全过程。

本书适合前后端开发人员和全栈工程师阅读,也适合高等院校和培训学校相关专业的师生教学参考。

本书封面贴有清华大学出版社防伪标签,无标签者不得销售。
版权所有,侵权必究。举报:010-62782989,beiqinquan@tup.tsinghua.edu.cn。

**图书在版编目(CIP)数据**

Spring Boot+Spring Cloud+Vue+Element 项目实战:手把手教你开发权限管理系统/徐丽健著.
—北京:清华大学出版社,2019(2021.5重印)
ISBN 978-7-302-52870-8

Ⅰ. ①S… Ⅱ. ①徐… Ⅲ. ①JAVA 语言—程序设计 Ⅳ. ①TP312.8

中国版本图书馆 CIP 数据核字(2019)第 083026 号

**责任编辑:**夏毓彦
**封面设计:**王　翔
**责任校对:**闫秀华
**责任印制:**丛怀宇

出版发行:清华大学出版社
　　　网　　址:http://www.tup.com.cn,http://www.wqbook.com
　　　地　　址:北京清华大学学研大厦 A 座　　　邮　编:100084
　　　社 总 机:010-62770175　　　邮　购:010-62786544
　　　投稿与读者服务:010-62776969,c-service@tup.tsinghua.edu.cn
　　　质量反馈:010-62772015,zhiliang@tup.tsinghua.edu.cn
印 装 者:三河市铭诚印务有限公司
经　　销:全国新华书店
开　　本:190mm×260mm　　　印　张:19.75　　　字　数:506 千字
版　　次:2019 年 6 月第 1 版　　　印　次:2021 年 5 月第 6 次印刷
定　　价:69.00 元

产品编号:083707-01

# 前　　言

纵观当今 Web 开发领域，优秀的后端开发语言虽有不少，但是 Java 依然独占鳌头，连续多年占据了行业的半壁江山，特别是随着 Spring Boot 和 Spring Cloud 的诞生和流行，集智慧于大成的 Spring 技术体系成为行业开发的首选之一。在前端领域，也是各种框架齐出，技术更新日新月异，在众多的技术和框架中，Vue.js、React 和 Angular.js 算是当前核心框架中的佼佼者，各自占有不少市场份额　。市场代表需求，技术代表能力。显而易见，在当今开发领域中，谁能更好地掌握这些主流开发技术，谁就能在跟别人竞争的时候多一些筹码，谁就能获得更好的就业机会、薪资报酬和发展空间。

如何能更好地掌握行业技术呢？抱着技术书籍埋头苦读？当然不能死记硬背，我们这样的开发人员，除了要掌握基础理论，最重要的还是要多实践，实践出真知嘛，大家都知道。要想更好地掌握开发技术和知识，就要进入项目多写代码，当然，对于大多数人来说，最好的成长方式，就是能够进入优秀的项目，跟着优秀的前辈，产出优秀的代码。然而现实是，很多人并不能够进入优秀的项目，也无法跟着优秀的前辈学习优秀的代码。他们苦于想要入门而又找不到门道，想要成长而又找不到方向，往往一不小心就在学习的路上混沌迷茫，不知所措，遂而丧失了信心，萌生了怯意。

## 关于本书

本书为广大开发者量身打造，从项目实践出发，选用当前各种主流的技术，手把手、心贴心地带着读者从零开始，一步一步地实现一个完整的后台权限管理系统。通过整个管理系统的开发和实践，让读者在学成之后能够熟悉和掌握当前的一些主流技术和方向，且在后续的工作中拥有自主搭建开发环境和完成整个系统开发的能力。后台权限管理系统是各种业务系统的基础配备模块之一，且整个大业务系统中的其他系统大多都要依赖权限系统模块，所以权限管理系统在整个业务系统中的重要性就不言而喻了。

本书的示例系统称为 Mango 权限管理系统，诞生于本教材实践项目。Mango 采用前后端分离架构，前端采用 Vue.js 作为核心框架，并使用同样非常流行的 Element 作为 UI 框架。前端开发基于 NPM 环境，使用 Visual Studio Code 作为 IDE 编写代码。前端使用 Mock 可以模拟后台接口数据，可以在没有后台的情况下使用大部分功能，所以也适合不会部署后端的开发人员学习和使用。后端则采用 Spring Boot + Spring Security + Spring Cloud + MyBatis 的主体架构，基于 Java 环境采用 Eclipse 开发，使用 Maven 工具构建，支持使用 Swagger 进行后台接口测试。总而言之，Mango 是一个基于 Spring Boot、Spring Cloud、Vue.js 、Element UI 实现，采用前后端分离架构的权限管理系统，也是一款采用当前主流技术实现的界面优雅、架构优良、代码简洁、注释完善、基础功能相对完整的 Java 快速开发平台。读者可以以此为范例从中学习和汲取技术知识，也可以基于此系统开发和实现具体生产项目。

本人知识有限，经验尚浅，书中若有谬处，烦请指正，不胜感激。回首当年，我也曾为如何入门而困扰，也因难以进步而迷茫，聊著此书，只为可以帮助更多的人在学习和开发中寻得门道、取得进步和成长，果有人能因此书而获益，那就是功德无量了。本书适用于业界前后端开发人员和全栈工程师以及广大想要学习和掌握前后端技术的人员，特别适合那些想要快速提升项目实践经验，熟悉和掌握架构开发整个业务系统能力的广大开发者。读者在学习和开发的过程中若有什么疑问，欢迎通过电子邮件提问或群聊咨询。

## 关于作者

徐丽健，毕业于广东金融学院，拥有多年 Java 开发和系统架构经验，开源技术爱好者和开源代码贡献者，闲暇之余爱写文字，博客园不知名技术博主。现在国内任科技企业的系统架构师一职，主持企业软件的系统架构和产品研发工作。

作者邮箱：514045152@qq.com
技术博客：https://www.cnblogs.com/xifengxiaoma/
开源社区：https://gitee.com/liuge1988
本书 QQ 技术交流群：429854222

## 代码下载

本书示例代码需要从 gitee 上下载，地址如下：

https://gitee.com/liuge1988/mango-platform

如果下载有问题，或者对本书有疑问和建议，请加入本书 QQ 技术交流群沟通。

著 者
2019 年 4 月

# 目　　录

## 第一篇　系统介绍篇

### 第1章　权限管理系统介绍 ·················································································· 3
#### 1.1　系统简介 ······································································································· 3
#### 1.2　系统架构 ······································································································· 4
##### 1.2.1　前端架构 ······························································································ 4
##### 1.2.2　后端架构 ······························································································ 4
#### 1.3　系统界面 ······································································································· 5
##### 1.3.1　登录页面 ······························································································ 5
##### 1.3.2　系统主页 ······························································································ 5
##### 1.3.3　用户管理 ······························································································ 6
##### 1.3.4　机构管理 ······························································································ 6
##### 1.3.5　角色管理 ······························································································ 7
##### 1.3.6　菜单管理 ······························································································ 7
##### 1.3.7　字典管理 ······························································································ 8
##### 1.3.8　系统配置 ······························································································ 8
##### 1.3.9　登录日志 ······························································································ 8
##### 1.3.10　操作日志 ···························································································· 9
##### 1.3.11　注册中心 ···························································································· 9
##### 1.3.12　接口文档 ···························································································· 9
##### 1.3.13　数据监控 ·························································································· 10
##### 1.3.14　服务监控 ·························································································· 11
##### 1.3.15　备份还原 ·························································································· 11
##### 1.3.16　主题切换 ·························································································· 11
### 第2章　安装指南 ······························································································ 13
#### 2.1　前端安装指南 ······························································································ 13
##### 2.1.1　开发环境 ···························································································· 13
##### 2.1.2　技术选型 ···························································································· 13
##### 2.1.3　项目结构 ···························································································· 13
##### 2.1.4　编译运行 ···························································································· 14
#### 2.2　后端安装指南 ······························································································ 14
##### 2.2.1　开发环境 ···························································································· 14
##### 2.2.2　技术选型 ···························································································· 15

| | 2.2.3 项目结构 | 15 |
|---|---|---|
| | 2.2.4 编译运行 | 15 |

## 第 3 章　关键技术 ································································································ 17

### 3.1 Spring Boot ································································································ 17
### 3.2 Spring Cloud ······························································································ 17
　　3.2.1 Spring Cloud 简介 ················································································ 17
　　3.2.2 Spring Cloud 架构 ················································································ 17
　　3.2.3 Spring Cloud 组件 ················································································ 18
　　3.2.4 参考教程 ···························································································· 19
### 3.3 Spring Security ·························································································· 19
### 3.4 MyBatis ···································································································· 19
### 3.5 Vue.js ······································································································ 19
### 3.6 Element ···································································································· 20

## 第二篇　后端实现篇

## 第 4 章　数据库设计 ···································································································· 23

### 4.1 数据库表设计 ······························································································ 23
### 4.2 数据库表关系 ······························································································ 23
### 4.3 数据库表结构 ······························································································ 24
　　4.3.1 用户表（sys_user） ············································································ 24
　　4.3.2 角色表（sys_role） ············································································ 25
　　4.3.3 机构表（sys_dept） ············································································ 25
　　4.3.4 菜单表（sys_menu） ·········································································· 26
　　4.3.5 用户角色表（sys_user_role） ································································ 26
　　4.3.6 角色菜单表（sys_role_menu） ······························································ 26
　　4.3.7 角色机构表（sys_role_dept） ································································ 27
　　4.3.8 字典表（sys_dict） ············································································ 27
　　4.3.9 配置表（sys_config） ·········································································· 28
　　4.3.10 操作日志表（sys_log） ······································································· 28
　　4.3.11 登录日志表（sys_login_log） ······························································· 28

## 第 5 章　搭建开发环境 ································································································ 30

### 5.1 开发环境准备 ······························································································ 30
　　5.1.1 安装 JDK 环境 ···················································································· 30
　　5.1.2 安装 Eclipse 开发工具 ·········································································· 30
　　5.1.3 安装 MySQL 数据库 ············································································ 30
　　5.1.4 安装 Maven 构建工具 ·········································································· 31
### 5.2 生成项目模板 ······························································································ 31
### 5.3 导入 Maven 项目 ························································································· 31

5.4 编译打包运行 ··································································································· 34
　　5.4.1 编译打包 ······························································································· 34
　　5.4.2 启动应用 ······························································································· 35
　　5.4.3 修改启动端口 ························································································ 35
　　5.4.4 自定义 Banner ······················································································· 35
　　5.4.5 接口测试 ······························································································· 37

# 第 6 章　集成 Swagger 文档 ··························································································· 38

6.1 添加依赖 ········································································································· 38
6.2 配置类 ············································································································ 39
6.3 页面测试 ········································································································· 39

# 第 7 章　集成 MyBatis 框架 ··························································································· 41

7.1 添加依赖 ········································································································· 41
7.2 添加配置 ········································································································· 42
　　7.2.1 添加 MyBatis 配置 ·················································································· 42
　　7.2.2 添加数据源配置 ····················································································· 42
　　7.2.3 修改启动类 ···························································································· 43
7.3 生成 MyBatis 模块 ··························································································· 43
7.4 编写服务接口 ··································································································· 44
7.5 配置打包资源 ··································································································· 46
7.6 编译运行测试 ··································································································· 47

# 第 8 章　集成 Druid 数据源 ··························································································· 48

8.1 Druid 介绍 ······································································································· 48
8.2 添加依赖 ········································································································· 49
8.3 添加配置 ········································································································· 49
8.4 配置 Servlet 和 Filter ························································································ 51
8.5 编译运行 ········································································································· 53
8.6 查看监控 ········································································································· 54
　　8.6.1 登录界面 ······························································································· 54
　　8.6.2 监控首页 ······························································································· 55
　　8.6.3 数据源 ·································································································· 55
　　8.6.4 SQL 监控 ······························································································· 55

# 第 9 章　跨域解决方案 ································································································· 57

9.1 什么是跨域 ······································································································ 57
9.2 CORS 技术 ······································································································ 57
　　9.2.1 简单请求 ······························································································· 57
　　9.2.2 非简单请求 ···························································································· 58
9.3 CORS 实现 ······································································································ 59

# 第 10 章 业务功能实现 ································································ 60

## 10.1 工程结构规划 ································································ 60
### 10.1.1 mango-admin ································································ 60
### 10.1.2 mango-common ································································ 62
### 10.1.3 mango-core ································································ 63
### 10.1.4 mango-pom ································································ 63
### 10.1.5 打包测试 ································································ 63

## 10.2 业务代码封装 ································································ 65
### 10.2.1 通用 CURD 接口 ································································ 65
### 10.2.2 分页请求封装 ································································ 66
### 10.2.3 分页结果封装 ································································ 67
### 10.2.4 分页助手封装 ································································ 68
### 10.2.5 HTTP 结果封装 ································································ 69

## 10.3 MyBatis 分页查询 ································································ 70
### 10.3.1 添加依赖 ································································ 70
### 10.3.2 添加配置 ································································ 71
### 10.3.3 分页代码 ································································ 71
### 10.3.4 接口测试 ································································ 73

## 10.4 业务功能开发 ································································ 74
### 10.4.1 编写 DAO 接口 ································································ 74
### 10.4.2 编写映射文件 ································································ 74
### 10.4.3 编写服务接口 ································································ 75
### 10.4.4 编写服务实现 ································································ 75
### 10.4.5 编写控制器 ································································ 77

## 10.5 业务接口汇总 ································································ 78
### 10.5.1 用户管理 ································································ 78
### 10.5.2 机构管理 ································································ 79
### 10.5.3 角色管理 ································································ 79
### 10.5.4 菜单管理 ································································ 80
### 10.5.5 字典管理 ································································ 81
### 10.5.6 系统配置 ································································ 82
### 10.5.7 登录日志 ································································ 82
### 10.5.8 操作日志 ································································ 83

## 10.6 导出 Excel 报表 ································································ 83
### 10.6.1 添加依赖 ································································ 83
### 10.6.2 编写服务接口 ································································ 84
### 10.6.3 编写服务实现 ································································ 84
### 10.6.4 编写控制器 ································································ 85
### 10.6.5 工具类代码 ································································ 86
### 10.6.6 接口测试 ································································ 87

# 第 11 章 登录流程实现 ·················································································· 89

## 11.1 登录验证码 ························································································ 89
### 11.1.1 添加依赖 ················································································· 89
### 11.1.2 添加配置 ················································································· 89
### 11.1.3 生成代码 ················································································· 90
### 11.1.4 接口测试 ················································································· 91

## 11.2 Spring Security ··················································································· 91
### 11.2.1 添加依赖 ················································································· 91
### 11.2.2 添加配置 ················································································· 92
### 11.2.3 登录认证过滤器 ········································································ 93
### 11.2.4 身份验证组件 ··········································································· 96
### 11.2.5 认证信息查询 ··········································································· 97
### 11.2.6 添加权限注解 ··········································································· 99
### 11.2.7 Swagger 添加令牌参数 ······························································ 100

## 11.3 登录接口实现 ···················································································· 101
## 11.4 接口测试 ·························································································· 105
## 11.5 Spring Security 执行流程剖析 ································································ 108

# 第 12 章 数据备份还原 ············································································· 109

## 12.1 新建工程 ·························································································· 109
## 12.2 添加依赖 ·························································································· 109
## 12.3 添加配置 ·························································································· 110
## 12.4 自定 Banner ····················································································· 111
## 12.5 启动类 ···························································································· 111
## 12.6 跨域配置 ························································································· 111
## 12.7 Swagger 配置 ··················································································· 112
## 12.8 数据源属性 ······················································································ 112
## 12.9 备份还原接口 ··················································································· 113
## 12.10 备份还原实现 ·················································································· 114
## 12.11 备份还原逻辑 ·················································································· 114
### 12.11.1 数据备份服务 ········································································ 115
### 12.11.2 数据还原服务 ········································································ 116
## 12.12 备份还原控制器 ··············································································· 117
### 12.12.1 数据备份接口 ········································································ 117
### 12.12.2 数据还原接口 ········································································ 117
### 12.12.3 查找备份接口 ········································································ 118
### 12.12.4 删除备份接口 ········································································ 119
## 12.13 接口测试 ························································································ 119

## 第 13 章　系统服务监控 ......123

- 13.1　新建工程 ......123
- 13.2　添加依赖 ......123
- 13.3　添加配置 ......124
- 13.4　自定义 Banner ......124
- 13.5　启动类 ......125
- 13.6　启动服务端 ......125
- 13.7　监控客户端 ......125
- 13.8　启动客户端 ......126

## 第 14 章　注册中心（Consul）......128

- 14.1　什么是 Consul ......128
- 14.2　Consul 安装 ......128
- 14.3　monitor 改造 ......129
  - 14.3.1　添加依赖 ......129
  - 14.3.2　配置文件 ......130
  - 14.3.3　启动类 ......130
  - 14.3.4　测试效果 ......131
- 14.4　backup 改造 ......132
  - 14.4.1　添加依赖 ......132
  - 14.4.2　配置文件 ......132
  - 14.4.3　启动类 ......133
  - 14.4.4　测试效果 ......134
- 14.5　admin 改造 ......134
  - 14.5.1　添加依赖 ......134
  - 14.5.2　配置文件 ......135
  - 14.5.3　启动类 ......136
  - 14.5.4　测试效果 ......137

## 第 15 章　服务消费（Ribbon、Feign）......138

- 15.1　技术背景 ......138
- 15.2　服务提供者 ......138
  - 15.2.1　新建项目 ......138
  - 15.2.2　配置文件 ......139
  - 15.2.3　启动类 ......140
  - 15.2.4　自定义 Banner ......140
  - 15.2.5　添加控制器 ......140
- 15.3　服务消费者 ......142
  - 15.3.1　新建项目 ......142
  - 15.3.2　添加配置 ......143
  - 15.3.3　启动类 ......144

  15.3.4 自定义 Banner ································································································· 144
  15.3.5 服务消费 ······································································································ 144
  15.3.6 负载均衡器（Ribbon）··················································································· 147
  15.3.7 修改启动类 ·································································································· 148
  15.3.8 添加服务 ······································································································ 149
  15.3.9 页面测试 ······································································································ 149
  15.3.10 负载策略 ···································································································· 149
 15.4 服务消费（Feign）································································································ 150
  15.4.1 添加依赖 ······································································································ 150
  15.4.2 启动类 ········································································································· 150
  15.4.3 添加 Feign 接口 ···························································································· 151
  15.4.4 添加控制器 ·································································································· 151
  15.4.5 页面测试 ······································································································ 152

## 第 16 章 服务熔断（Hystrix、Turbine）································································ 153

 16.1 雪崩效应 ·············································································································· 153
 16.2 熔断器（CircuitBreaker）······················································································ 153
 16.3 Hystrix 特性 ········································································································· 153
  16.3.1 断路器机制 ·································································································· 153
  16.3.2 fallback ······································································································· 154
  16.3.3 资源隔离 ······································································································ 154
 16.4 Feign Hystrix 154
  16.4.1 修改配置 ······································································································ 154
  16.4.2 创建回调类 ·································································································· 155
  16.4.3 页面测试 ······································································································ 155
 16.5 Hystrix Dashboard 156
  16.5.1 添加依赖 ······································································································ 156
  16.5.2 启动类 ········································································································· 157
  16.5.3 自定义 Banner ······························································································ 157
  16.5.4 配置文件 ······································································································ 158
  16.5.5 配置监控路径 ······························································································· 158
  16.5.6 页面测试 ······································································································ 159
 16.6 Spring Cloud Turbine 161
  16.6.1 添加依赖 ······································································································ 161
  16.6.2 启动类 ········································································································· 162
  16.6.3 配置文件 ······································································································ 162
  16.6.4 测试效果 ······································································································ 163

## 第 17 章 服务网关（Zuul）······················································································· 164

 17.1 技术背景 ·············································································································· 164
 17.2 Spring Cloud Zuul ································································································ 164

IX

17.3 Zuul 工作机制 ·············· 165
   17.3.1 过滤器机制 ·············· 165
   17.3.2 过滤器的生命周期 ·············· 165
   17.3.3 禁用指定的 Filter ·············· 167
17.4 实现案例 ·············· 167
   17.4.1 新建工程 ·············· 167
   17.4.2 添加依赖 ·············· 168
   17.4.3 启动类 ·············· 168
   17.4.4 配置文件 ·············· 169
   17.4.5 页面测试 ·············· 169
   17.4.6 配置接口前缀 ·············· 170
   17.4.7 默认路由规则 ·············· 170
   17.4.8 路由熔断 ·············· 171
   17.4.9 自定义 Filter ·············· 172

## 第 18 章 链路追踪（Sleuth、ZipKin）·············· 174

18.1 技术背景 ·············· 174
18.2 ZipKin ·············· 174
18.3 Spring Cloud Sleuth ·············· 174
18.4 实现案例 ·············· 175
   18.4.1 下载镜像 ·············· 175
   18.4.2 编写启动文件 ·············· 175
   18.4.3 启动服务 ·············· 176
   18.4.4 添加依赖 ·············· 177
   18.4.5 配置文件 ·············· 178
   18.4.6 页面测试 ·············· 178

## 第 19 章 配置中心（Config、Bus）·············· 180

19.1 技术背景 ·············· 180
19.2 Spring Cloud Config ·············· 180
19.3 实现案例 ·············· 181
   19.3.1 准备配置文件 ·············· 181
   19.3.2 服务端实现 ·············· 181
   19.3.3 客户端实现 ·············· 185
   19.3.4 Refresh 机制 ·············· 188
   19.3.5 Spring Cloud Bus ·············· 191

# 第三篇 前端实现篇

## 第 20 章 搭建开发环境 ·············· 201

20.1 技术基础 ·············· 201

## 20.2 开发环境 ········································································································ 201
### 20.2.1 Visual Studio Code ················································································ 201
### 20.2.2 Node JS ······························································································· 202
### 20.2.3 安装 webpack ······················································································· 203
### 20.2.4 安装 vue-cli ·························································································· 203
### 20.2.5 淘宝镜像 ······························································································· 203
### 20.2.6 安装 Yarn ···························································································· 203
## 20.3 创建项目 ········································································································ 204
### 20.3.1 生成项目 ······························································································· 204
### 20.3.2 安装依赖 ······························································································· 205
### 20.3.3 启动运行 ······························································································· 206

# 第 21 章 前端项目案例 ······························································································ 207
## 21.1 导入项目 ········································································································ 207
## 21.2 安装 Element ·································································································· 207
### 21.2.1 安装依赖 ······························································································· 207
### 21.2.2 导入项目 ······························································································· 208
## 21.3 页面路由 ········································································································ 210
### 21.3.1 添加页面 ······························································································· 210
### 21.3.2 配置路由 ······························································································· 210
## 21.4 安装 SCSS ······································································································ 212
### 21.4.1 安装依赖 ······························································································· 212
### 21.4.2 添加配置 ······························································································· 212
### 21.4.3 如何使用 ······························································································· 212
### 21.4.4 页面测试 ······························································································· 212
## 21.5 安装 axios ······································································································ 213
### 21.5.1 安装依赖 ······························································································· 213
### 21.5.2 编写代码 ······························································································· 213
### 21.5.3 页面测试 ······························································································· 214
## 21.6 安装 Mock.js ···································································································· 214
### 21.6.1 安装依赖 ······························································································· 215
### 21.6.2 编写代码 ······························································································· 215
### 21.6.3 页面测试 ······························································································· 216

# 第 22 章 工具模块封装 ······························································································ 217
## 22.1 封装 axios 模块 ································································································ 217
### 22.1.1 封装背景 ······························································································· 217
### 22.1.2 封装要点 ······························································································· 217
### 22.1.3 文件结构 ······························································································· 217
### 22.1.4 代码说明 ······························································································· 218
### 22.1.5 安装 js-cookie ······················································································· 222

   22.1.6 测试案例 ··········································································································· 223
 22.2 封装 mock 模块 ····································································································· 225
   22.2.1 文件结构 ··········································································································· 225
   22.2.2 登录界面 ··········································································································· 228
   22.2.3 主页界面 ··········································································································· 229
   22.2.4 页面测试 ··········································································································· 229

## 第 23 章 第三方图标库 ·············································································································· 230

 23.1 使用第三方图标库 ································································································· 230
 23.2 Font Awesome ········································································································ 230
   23.2.1 安装依赖 ··········································································································· 230
   23.2.2 项目引入 ··········································································································· 230
   23.2.3 页面使用 ··········································································································· 231
   23.2.4 页面测试 ··········································································································· 231

## 第 24 章 多语言国际化 ·············································································································· 232

 24.1 安装依赖 ················································································································ 232
 24.2 添加配置 ················································································································ 232
 24.3 字符引用 ················································································································ 234
 24.4 页面测试 ················································································································ 235

## 第 25 章 登录流程完善 ·············································································································· 236

 25.1 登录界面 ················································································································ 236
   25.1.1 界面设计 ··········································································································· 236
   25.1.2 关键代码 ··········································································································· 236
 25.2 主页面 ···················································································································· 237
   25.2.1 界面设计 ··········································································································· 237
   25.2.2 关键代码 ··········································································································· 237
 25.3 页面测试 ················································································································ 240

## 第 26 章 管理应用状态 ·············································································································· 241

 26.1 安装依赖 ················································································································ 241
 26.2 添加 store ·············································································································· 241
   26.2.1 index.js ············································································································· 242
   26.2.2 app.js ··············································································································· 242
 26.3 引入 Store ············································································································· 243
 26.4 使用 Store ············································································································· 243
 26.5 收缩组件 ················································································································ 244
   26.5.1 文件结构 ··········································································································· 244
   26.5.2 关键代码 ··········································································································· 244
 26.6 页面测试 ················································································································ 245

# 第 27 章　头部功能组件 ··········································· 247

## 27.1　主题切换组件 ··········································· 247
### 27.1.1　编写组件 ··········································· 247
### 27.1.2　页面测试 ··········································· 250
## 27.2　语言切换组件 ··········································· 250
### 27.2.1　编写组件 ··········································· 250
### 27.2.2　页面测试 ··········································· 251
## 27.3　用户信息面板 ··········································· 252
### 27.3.1　编写组件 ··········································· 252
### 27.3.2　页面测试 ··········································· 253
## 27.4　系统通知面板 ··········································· 254
### 27.4.1　编写组件 ··········································· 254
### 27.4.2　页面测试 ··········································· 255
## 27.5　用户私信面板 ··········································· 255
### 27.5.1　编写组件 ··········································· 255
### 27.5.2　页面测试 ··········································· 256

# 第 28 章　动态加载菜单 ··········································· 258

## 28.1　添加 Store ··········································· 258
## 28.2　登录页面 ··········································· 259
## 28.3　导航守卫 ··········································· 259
## 28.4　导航树组件 ··········································· 262
## 28.5　页面测试 ··········································· 263

# 第 29 章　页面权限控制 ··········································· 264

## 29.1　权限控制方案 ··········································· 264
### 29.1.1　菜单类型 ··········································· 264
### 29.1.2　权限标识 ··········································· 264
### 29.1.3　菜单表结构 ··········································· 264
## 29.2　导航菜单实现思路 ··········································· 265
### 29.2.1　用户登录系统 ··········································· 265
### 29.2.2　根据用户加载导航菜单 ··········································· 265
### 29.2.3　导航栏读取菜单树 ··········································· 265
## 29.3　页面按钮实现思路 ··········································· 265
### 29.3.1　用户登录系统 ··········································· 265
### 29.3.2　加载权限标识 ··········································· 266
### 29.3.3　页面按钮控制 ··········································· 266
## 29.4　权限控制实现 ··········································· 266
### 29.4.1　导航菜单权限 ··········································· 266
### 29.4.2　页面按钮权限 ··········································· 267
## 29.5　标签页功能 ··········································· 270

29.6 系统介绍页 ····· 273
29.7 页面测试 ····· 274

# 第30章 功能管理模块 ····· 276

30.1 字典管理 ····· 276
 30.1.1 关键代码 ····· 276
 30.1.2 页面截图 ····· 279
30.2 角色管理 ····· 279
 30.2.1 关键代码 ····· 279
 30.2.2 页面截图 ····· 281
30.3 菜单管理 ····· 281
 30.3.1 表格列组件 ····· 282
 30.3.2 创建表格树 ····· 283
 30.3.3 页面截图 ····· 284

# 第31章 嵌套外部网页 ····· 285

31.1 需求背景 ····· 285
31.2 实现原理 ····· 285
31.3 代码实现 ····· 285
 31.3.1 确定菜单URL ····· 285
 31.3.2 创建嵌套组件 ····· 287
 31.3.3 绑定嵌套组件 ····· 288
 31.3.4 菜单路由跳转 ····· 290
31.4 页面测试 ····· 290

# 第32章 数据备份还原 ····· 293

32.1 需求背景 ····· 293
32.2 后台接口 ····· 293
32.3 备份页面 ····· 294
32.4 页面引用 ····· 296
32.5 页面测试 ····· 298

# 第一篇

# 系统介绍篇

本篇内容主要包括系统介绍、安装指南和关键技术 3 个章节内容。

第 1 章 权限管理系统介绍，从系统功能、系统架构和系统界面 3 个方面出发，分别进行解读和描述，让读者在开始学习和开发前，能够对本书内容和项目有一个大致的印象。

第 2 章 安装指南，分为前后端安装指南两个部分，通过完整详细的项目安装运行步骤，帮助读者快速在本地搭建起开发环境，快速进入源码学习和项目实践。

第 3 章 关键技术，为读者介绍 Mango 权限管理系统开发中所涉及的主要技术，让读者对涉及的相关技术有一个初步的认识，并引导读者进行更为深入的学习。

# 第 1 章 权限管理系统介绍

本章分为权限管理系统介绍、系统架构和系统界面 3 节,针对基于本书实现的 Mango(本书示例项目名)权限管理系统,分别从系统功能、系统架构和系统界面 3 个方面进行相对整体的介绍,让读者对 Mango 系统以及本书涉及的相关技术有一个大致的印象和了解,以便在后续的阅读中可以结合相关知识和项目实践逐步深入学习和开发。系统简介罗列主要的系统功能,系统架构分别对前后端的架构绘图进行描述,系统界面通过系统功能界面截图并配合简要描述的方式帮助读者更好地了解系统拥有的功能和模块。

## 1.1 系统简介

Mango 后台权限管理系统是基于 Spring Boot、Spring Cloud、Vue.js 、Element UI 等主流前后端技术,采用前后端分离架构实现的权限管理系统,也是一款采用当前主流技术实现的界面优雅、架构优良、代码简洁、注释完善、基础功能相对完整的 Java EE 快速开发平台,前后端开发人员都可以以此为范例从中学习和汲取技术知识,也可以基于此系统开发和实现具体生产项目。

Mango 实现的主要功能包括:

- 系统登录:系统用户登录,系统登录认证(token 方式)。
- 用户管理:新建用户,修改用户,删除用户,查询用户。
- 机构管理:新建机构,修改机构,删除机构,查询机构。
- 角色管理:新建角色,修改角色,删除角色,查询角色。
- 菜单管理:新建菜单,修改菜单,删除菜单,查询菜单。
- 字典管理:新建字典,修改字典,删除字典,查询字典。
- 配置管理:新建配置,修改配置,删除配置,查询配置。
- 登录日志:记录用户的登录日志,查看系统登录日志记录。
- 操作日志:记录用户的操作日志,查看系统操作日志记录。
- 在线用户:根据用户的登录状态,查看统计当前在线用户。
- 数据监控:定制 Druid 信息,提供简洁有效的 SQL 数据监控。
- 聚合文档:定制 Swagger 文档,提供简洁美观的 API 文档。

- 备份还原：系统数据备份还原，一键恢复系统初始化数据。
- 主题切换：支持主题切换，自定主题颜色，实现一键换肤。
- 服务治理：集成 Consul 注册中心，实现服务的注册和发现。
- 服务监控：集成 Spring Boot Admin，实现全方位的服务监控。
- 服务消费：集成 Ribbon、Feign，实现服务调用和负载均衡。
- 服务熔断：集成 Hystrix、Turbine，实现服务的熔断和监控。
- 服务网关：集成 Spring Cloud Zuul，实现统一 API 服务网关。
- 链路追踪：集成 Sleuth、ZipKin，实现服务分布式链路追踪。
- 配置中心：集成 Cloud Config 和 Bus，实现分布式配置中心。

这里简单地罗列了相关的系统功能，后续篇幅会给出各功能的系统界面说明。

## 1.2 系统架构

本系统采用前后端分离架构实现，前后端通过 JSON 格式进行交互，前后端皆可分开独立部署。前端支持开启 Mock 模拟接口数据，可以避免对后台接口开发进度的依赖；后台支持使用 Swagger 进行接口测试，同样可以避免对前端页面开发进度的依赖。

### 1.2.1 前端架构

前端架构比较简单，核心框架使用当前主流的 Vue.js，UI 使用饿了么开源的 Element，前后端交互使用了 axios，使用 Mock 模拟接口数据。

前端架构如图 1-1 所示。

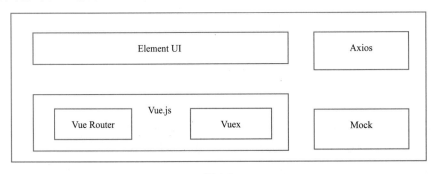

图 1-1

### 1.2.2 后端架构

后端架构使用 Spring Boot + Spring Security + Spring Cloud + MyBatis 的主体架构，除此之外，选择 Consul 注册中心，使用 Maven 构建工具、MySQL 数据库等。

后端架构如图 1-2 所示。

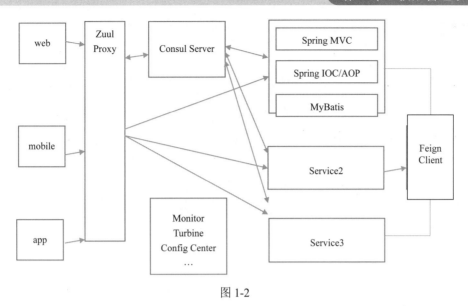

图 1-2

## 1.3 系统界面

### 1.3.1 登录页面

系统登录界面如图 1-3 所示,因为验证码是通过后台生成的,所以显示验证码需要后台的支持。

图 1-3

### 1.3.2 系统主页

系统主页主要是系统介绍内容,如图 1-4 所示。

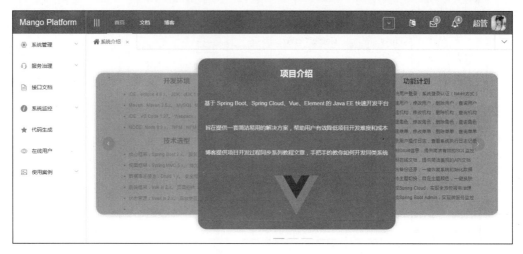

图 1-4

## 1.3.3 用户管理

用户管理支持用户的增、删、改、查功能，用户可以配置角色和机构。另外，在工具栏提供了表格列过滤和导出 Excel 表格的实现范例，如图 1-5 所示。

图 1-5

## 1.3.4 机构管理

机构管理支持机构的增、删、改、查功能，是树型层级结构，如图 1-6 所示。

图 1-6

## 1.3.5 角色管理

角色管理支持角色的增、删、改、查功能，支持给选定角色配置不同的菜单权限，如图 1-7 所示。

图 1-7

## 1.3.6 菜单管理

菜单管理支持菜单的增、删、改、查功能，是树型层级结构。菜单可以设置类型、图标、权限标识和菜单路径 URL 属性，如图 1-8 所示。

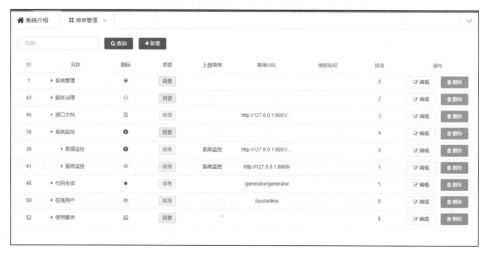

图 1-8

## 1.3.7 字典管理

字典管理支持数据字典的增、删、改、查功能，如图 1-9 所示。

图 1-9

## 1.3.8 系统配置

系统配置支持系统配置信息的增、删、改、查功能，如图 1-10 所示。

图 1-10

## 1.3.9 登录日志

登录日志支持系统登录日志的查询，如图 1-11 所示。

图 1-11

## 1.3.10 操作日志

操作日志支持系统用户操作日志的查询，如图 1-12 所示。

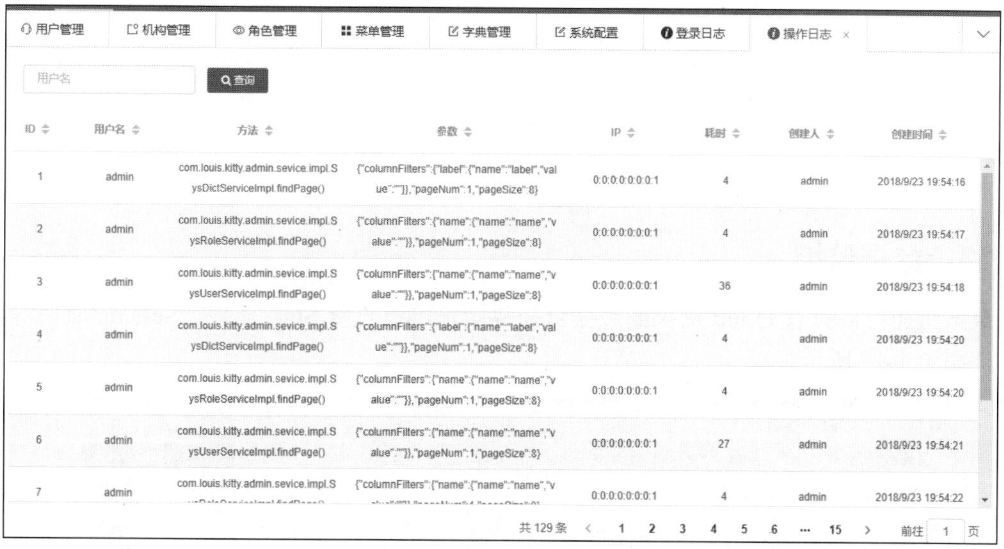

图 1-12

## 1.3.11 注册中心

注册中心可以查看 Consul 注册中心的服务页面，可以查看当前服务的注册情况和健康状况信息，如图 1-13 所示。

## 1.3.12 接口文档

接口文档可以查看 Swagger 集成的接口文档页面，如图 1-14 所示。

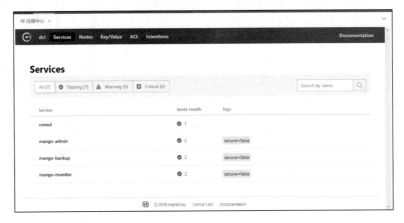

图 1-13

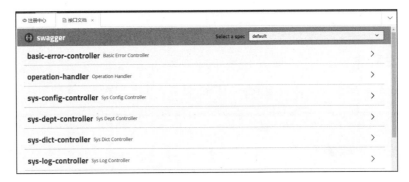

图 1-14

## 1.3.13 数据监控

数据监控可以查看 Druid 提供的数据监控界面，可以查看 SQL 监控、Session 监控等信息内容，如图 1-15 所示。

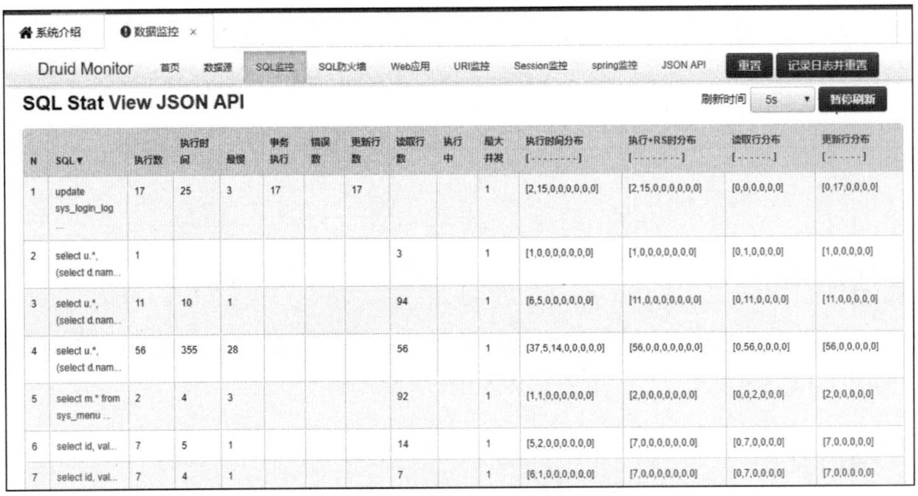

图 1-15

## 1.3.14 服务监控

服务监控可以查看 Spring Boot Admin 提供的 Spring Boot 应用监控页面，查看应用的 CPU、线程、内存和垃圾回收状态等信息，如图 1-16 所示。

图 1-16

## 1.3.15 备份还原

备份还原入口在用户信息面板，可以提供系统数据的备份和还原，包括备份创建、备份删除、备份查询和备份还原功能，如图 1-17 所示。

图 1-17

## 1.3.16 主题切换

主题切换支持通过主题切换器实现一键换肤的功能。

1. 绿色主题（如图1-18所示）

图1-18

2. 黑色主题（如图1-19所示）

图1-19

# 第 2 章 安装指南

本章节介绍如何本地安装运行 Mango 权限管理系统，着眼于实践，建议读者结合源码和书本内容逐步学习和掌握 Mango 系统的整个开发过程。因为是前后端分离项目，所以项目安装指南分为前端安装指南和后端安装指南两部分。

## 2.1 前端安装指南

本系统前端使用 Vue.js 和 Element 框架搭建，基于 NPM 环境开发，所以在开发之前，需要先安装 Node.JS，开发工具为 Visual Studio Code，当然读者也可以根据自己的喜好选择其他开发工具，比如说 WebStorm，具体开发环境的搭建请参考后续章节：前端实现篇◎搭建开发环境，这里着重说明以下现有源码的安装和运行。

### 2.1.1 开发环境

前端开发环境基于 NPM 环境，使用 VS Code 开发。

```
IDE : VS Code 1.27
NODE: Node 10.15.x
NPM : NPM 6.4.x
…
```

### 2.1.2 技术选型

前端技术主要使用 Vue.js 和 Element UI 框架。

```
前端框架: Vue 2.x
页面组件: Element 2.x
状态管理: Vuex 2.x
后台交互: axios 0.18.x
图标使用: Font Awesome 4.x
…
```

### 2.1.3 项目结构

前端项目结构如下：

```
mango-ui
-- build：项目编译相关模块，项目模板自动生成
-- config：项目配置相关模块，项目模板自动生成
-- src：项目源码模块，前端开发工作集中在此目录
    -- assets：图标、字体、国际化信息等静态信息
    -- components：组件库，对常用组件进行封装
    -- http：后台交互模块，统一后台接口请求 API
    -- i18n：国际化模块，使用 Vue i18n 进行国际化
    -- mock：Mock 模块，模拟接口调用并返回定制数据
    -- permission：权限控制模块，处理权限认证逻辑
    -- router：路由管理模块，负责页面各种路由配置
    -- store：状态管理模块，提供组件间状态共享
    -- utils：工具模块，提供一些通用的工具方法
    -- views：页面模块，主要放置各种页面视图组件
```

### 2.1.4 编译运行

编译运行步骤说明如下。

（1）获取源码。获取前端源码，整个前端只有一个工程 mango-ui，将其备份放置到本地目录。

（2）编译源码。在 mango-ui 目录下打开 CMD 终端，执行 npm install，下载和安装项目依赖包。

（3）启动系统。执行 npm run dev 命令，启动项目，启动之后通过 http://localhost:8080 访问。

（4）项目打包。执行 npm run build 命令，进行前端项目打包，打包完成之后会生成 dist 目录。

将生成的目录直接放置到如 Tomcat 之类的 Web 服务器，启动服务即可访问。

（5）Mock 开关。本系统采用前后端分离架构，前端若开启 Mock 模块，则可模拟大部分接口数据。

通过修改 src/mock/index.js 中的 openMock 变量，可以一键开启或关闭 Mock 功能。

（6）修改配置。如果想自定义端口（默认是 8080），可以修改 config/index.js 下的 port 属性。

后台接口和备份服务器地址配置在 src/utils/global.js，如有修改请做相应变更。

## 2.2 后端安装指南

### 2.2.1 开发环境

后端开发环境基于 Java 环境，使用 Eclipse 开发。

```
IDE : eclipse 4.x
JDK : JDK1.8.x
Maven : Maven 3.5.x
MySQL: MySQL 5.7.x
Consul: Consul 1.4.0
……
```

### 2.2.2 技术选型

后端技术主要使用 Spring Boot、Spring Cloud 和 MyBatis 框架。

```
核心框架：Spring Boot 2.x
服务治理：Spring Cloud Finchley
安全框架：Spring Security 5.x
视图框架：Spring MVC 5.x
持久层框架：MyBatis 3.x
数据库连接池：Druid 1.x
消息队列：RabbitMQ
接口文档：Swagger 2.9.x
日志管理：SLF4J、Log4j
……
```

### 2.2.3 项目结构

后端项目源码工程结构如下：

```
mango-common：公共代码模块，主要放置一些工具类
mango-core：封装业务模块，主要封装公共业务模块
mango-admin：后台管理模块，包含用户、角色、菜单管理等
mango-backup：系统数据备份还原模块，可选择独立部署
mango-monitor：系统监控服务端，监控 Spring Boot 应用
mango-producer：服务提供者示例，方便在此基础上搭建模块
mango-consumer：服务消费者示例，方便在此基础上搭建模块
mango-hystrix：服务熔断监控模块，收集汇总熔断统计信息
mango-zuul：API 服务网关模块，统一管理和转发外部调用请求
mango-config：配置中心服务端，生成 GIT 配置文件的访问接口
mango-consul：注册中心，安装说明目录，内附安装引导说明
mango-zipkin：链路追踪，安装说明目录，内附安装引导说明
config-repo：配置中心仓库，在 GIT 上统一存储系统配置文件
mango-pom：聚合模块，仅为简化打包，一键执行打包所有模块
```

### 2.2.4 编译运行

**1. 编译运行步骤**

（1）获取源码。获取后端源码，获取上面所列的所有项目结构，将其备份放置到本地目录。

（2）导入工程。使用 Eclipse 导入 Maven 项目，在此之前请确认已安装 JDK 和 Maven 工具。

（3）编译源码。找到 mango-pom 工程下的 pom.xml，执行 maven clean install 命令进行一键打包。一般来说不会有什么问题，如果打包失败，可以尝试按照优先级逐个编译。

（4）导入数据库。新建 mango 数据库，使用项目 sql 目录下的 mango.sql 脚本，导入初始化数据库。

修改 mango-admin 下 application.yml 中的数据源配置信息为自己的数据库配置。
修改 mango-backup 下 application.yml 中的数据源配置信息为自己的数据库配置。

（5）启动系统

- 基础必需模块（注册中心，mango-consul；服务监控，mango-monitor）

找到 mango-consul 工程，根据安装说明安装注册中心，执行 consul agent -dev 启动。
找到 mango-monitor 工程下的 MangoMonitorApplication，启动项目，开启服务监控。

- 权限管理模块（权限管理，mango-admin；备份还原，mango-backup）

找到 mango-admin 工程下的 MangoAdminApplication，启动项目，开启权限系统服务。
找到 mango-backup 工程下的 MangoBackupApplication.java，启动项目，开启备份服务。

- 其他示例模块（Spring Cloud 示例模块，作为开发模板和范例，根据需要启动）

以下为 Spring Cloud 体系各种功能的实现范例，可以根据需要启动，后续扩展开发也可以作为参考和模板使用，具体使用教程请参考本书后面 Spring Cloud 系列教程的章节，关于 Spring Cloud 体系的各种功能模块都有详细的讲解和完整的案例实现。

这些示例模块包括：

- mango-producer：服务提供者示例，演示服务提供者的实现。
- mango-consumer：服务消费者示例，演示服务消费者的实现。
- mango-hystrix：服务熔断监控模块，演示熔断监控功能的实现。
- mango-zuul：API 服务网关模块，演示 API 统一网关的实现。
- mango-config：配置中心服务端，演示分布式配置中心的实现。

2. 注意事项

（1）注册中心是基础服务，需要先安装 Consul，找到 mango-consul 工程，根据安装说明安装 Consul。

（2）如果需要链路追踪服务，需要安装 zipkin，找到 mango-zipkin 工程，根据安装说明安装 zipkin。

（3）如果需要配置中心服务，需要安装 rabbitMQ，找到 mango-config 工程，根据安装说明安装 rabbitMQ。

# 第 3 章 关键技术

## 3.1 Spring Boot

Spring Boot 是由 Pivotal 团队提供，设计用来简化新 Spring 应用的初始搭建和开发过程的开源框架。大家都知道，随着 Spring 体系越来越庞大，各种配置也是越来越繁杂，曾经有多少开发人员在埋头修改配置文件的时候对此嗤之以鼻，直到 Spring Boot 的诞生。Spring Boot 遵循约定优于配置的规则，使用特定的方式来进行配置，从而使开发人员不再需要定义各种样板化的配置，将开发人员从繁杂的配置文件中解放出来。

Spring Boot 其实并不是什么新的框架，它只是默认配置了很多框架的使用方式，就像 Maven 整合了所有的 jar 包一样，Spring Boot 整合了所有的框架。使用 Spring Boot 可以非常方便、快速地搭建我们的项目，使我们可以不用关心各框架之间的兼容性、版本适用性等问题，我们想使用什么东西，往往只需添加一个配置即可。另外，Spring Boot 内置了服务器，使得服务的测试和部署也变得非常方便，所以 Spring Boot 非常适合开发微服务。

官方教程：http://spring.io/projects/spring-boot/ 。

## 3.2 Spring Cloud

### 3.2.1 Spring Cloud 简介

Spring Cloud 是一个微服务框架。相比 Dubbo 等 RPC 框架，Spring Cloud 提供的是一整套分布式系统解决方案。它为微服务架构开发中所涉及的服务治理、服务熔断、智能路由、链路追踪、消息总线、配置管理、集群状态管理等操作都提供了一种简单的开发方式。

### 3.2.2 Spring Cloud 架构

Spring Cloud 的架构大致如图 3-1（来自互联网）所示。

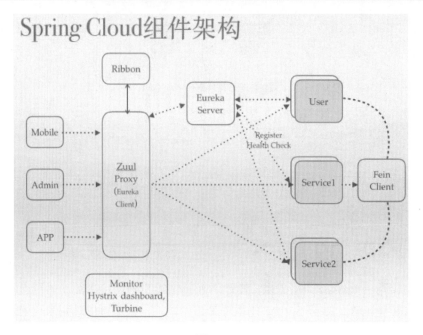

图 3-1

### 3.2.3 Spring Cloud 组件

Spring Cloud 是一个全套的框架，整个框架系统包含各种方方面面的功能组件，组件数量繁多，这里针对一些比较常用的组件进行说明。

- Netflix Eureka：一个基于 REST 服务的服务治理组件，包括服务注册中心、服务注册与服务发现机制的实现，目前仍是 Spring Cloud 的默认内置注册中心组件，不过自从 Eureka 2.0 停止开发的消息以后，有逐渐被 Consul 等注册中心替代的趋势。
- Netflix Hystrix：容错管理工具，实现断路器模式，通过控制服务的节点，从而对延迟和故障提供更强大的容错能力。
- Netflix Ribbon：实现客户端负载均衡的服务调用组件。
- Netflix Feign：基于 Ribbon 和 Hystrix 的声明式服务调用组件。
- Netflix Zuul：微服务网关，提供动态路由，访问过滤等服务。
- Spring Cloud Config：配置管理工具，支持使用 Git 存储配置内容，支持应用配置的外部化存储，支持客户端配置信息刷新、加解密配置内容等。
- Spring Cloud Bus：事件、消息总线，用于在集群（例如，配置变化事件）中传播状态变化，可与 Spring Cloud Config 联合实现热部署。
- Spring Cloud Sleuth：日志收集工具包，封装了 Dapper、ZipKin 和 HTrace 操作。
- Spring Cloud Consul：封装了 Consul 操作，Consul 是一个服务发现与配置工具，与 Docker 容器可以无缝集成。
- Spring Cloud Zookeeper：操作 Zookeeper 的工具包，用于使用 zookeeper 方式的服务注册和发现。

- Spring Cloud Stream：数据流操作开发包，封装了与 Redis、Rabbit、Kafka 等发送接收消息的接口。
- Spring Cloud Data Flow：大数据操作工具，通过命令行方式操作数据流。
- Spring Cloud Security：安全工具包，为你的应用程序添加安全控制，主要是指 OAuth2。

### 3.2.4 参考教程

提供一些 Spring Cloud 的参考资料：

- 官网教程：http://spring.io/projects/spring-cloud
- 博客教程：https://www.cnblogs.com/xifengxiaoma/p/9798330.html

## 3.3 Spring Security

Spring Security 是 Spring 社区的一个顶级项目，也是 Spring Boot 官方推荐使用的安全框架。Spring Security 是一个强大且支持高度自定义的安全框架，可以为系统在登录认证和访问授权两大核心安全方面提供强有力的保障。

- 官网教程：https://spring.io/projects/spring-security
- 博客教程：https://www.cnblogs.com/xifengxiaoma/p/10020960.html

## 3.4 MyBatis

MyBatis 是一款非常优秀的持久层框架，可以支持定制化 SQL、存储过程以及高级映射等高级特性。MyBatis 可以使用简单的 XML 或注解来配置和映射原生信息，将接口和 Java 的 POJOs 对象映射成数据库中的记录，使用 MyBatis 可以避免几乎所有的 JDBC 代码，无须手动设置参数和获取结果集。相比类似 Hibernate 这种完全 ORM 框架，MyBatis 其实只能算是 SQL Mapping 框架，两种方式各有优劣，MyBatis 虽然编写 Mapping 文件有点麻烦，但贵在灵活，故深受企业的喜爱，成为当前主流的持久层框架。

- 官网教程：http://www.mybatis.org/mybatis-3/zh/
- W3C 教程：https://www.w3cschool.cn/mybatis/

## 3.5 Vue.js

Vue.js 是当前前端领域主流的核心框架之一，特别深受国内用户的喜爱。

Vue.js（读音 /vju:/，类似于 view）是一套构建用户界面的渐进式框架。
Vue 只关注视图层，采用自底向上增量开发的设计方式，非常容易上手。
Vue 的目标是通过尽可能简单的 API 实现响应的数据绑定和组合的视图组件。

- 官网教程：https://cn.vuejs.org/v2/guide/
- 菜鸟教程：http://www.runoob.com/vue2/vue-tutorial.html

## 3.6 Element

Element 是国内饿了么开源的一套前端 UI 框架，提供了较为丰富的组件，界面简洁优雅，同样深受国内开发者的喜爱。Element 分别提供了 Vue.js、React 和 Angular 的实现，本书将使用 Vue.js 版本搭建 Mango 管理系统的界面。

- 官网教程：http://element.eleme.io/#/zh-CN

# 第二篇
# 后端实现篇

本篇内容为后端实现篇,全面细致地讲解了 Mango 权限管理系统的后端实现全过程。从零开始,逐步扩展,逐渐完善,手把手地教你如何利用 Spring Boot 和 Spring Cloud 构建微服务系统。

第 4 章 数据库设计,详细地阐述设计原则、表间关系和数据库表结构。

第 5 章 搭建开发环境,完整地阐述和示范后端开发环境的搭建和安装。

第 6 章 集成 Swagger 文档,阐述和实现如何集成 Swagger 并进行接口测试。

第 7 章 集成 MyBatis 框架,阐述和实现如何集成 MyBatis 进行数据库操作。

第 8 章 集成 Druid 数据源,阐述和实现如何集成 Druid 数据源和查看 SQL 监控。

第 9 章 跨域解决方案,阐述什么是跨域并提供 CORS 实现跨域的解决方案。

第 10 章 业务功能实现,对权限系统后台涉及的业务功能接口统一设计开发。

第 11 章 登录流程实现,讲解如何集成 Spring Security 并完善登录流程。

第 12 章 数据备份还原,讲解如何通过调用 MySQL 命令完成数据备份还原。

第 13 章 系统服务监控,讲解如何集成 Spring Boot Admin 实现服务监控。

第 14 章 注册中心(Consul),讲解如何安装 Consul 注册中心和服务客户端的注册。

第 15 章 服务消费(Ribbon、Feign),阐述和实现如何通过 Ribbon 和 Feign 进行服务消费。

第 16 章 服务熔断(Hystrix、Turbine),讲解如何集成 Hystrix 和 Turbine 进行服务熔断和监控。

第 17 章 服务网关(Zuul),阐述和示范如何通过 Zuul 实现智能路由,提供 API 网关。

第 18 章 链路追踪(Sleuth、ZipKin),讲解如何集成 Sleuth 和 ZipKin 进行服务调用的链路追踪。

第 19 章 配置中心(Config、Bus),讲解如何通过 Spring Cloud Config 实现分布式配置中心。

# 第 4 章 数据库设计

## 4.1 数据库表设计

在经过对权限管理系统做了细致的需求分析和功能分析之后，我们列出需要实现的需求：

- 系统登录：能够进行系统登录。
- 用户管理：提供用户管理界面。
- 机构管理：提供机构管理界面。
- 角色管理：提供角色管理界面。
- 菜单管理：提供菜单管理界面。
- 字典管理：提供字典管理界面。
- 登录日志：提供登录日志界面。
- 操作日志：提供操作日志界面。
- 数据监控：集成 Druid 数据监控。
- 服务监控：集成 Boot Admin 服务监控。
- 服务治理：集成 Spring Cloud 服务治理。

基于以上需求，结合各功能需求之间的直接关系，我们需要设计出权限管理系统的数据库表。

这里，我们的数据库设计遵循以下原则：

- 所有数据库表均采用长整型的"编号"字段作为表的主键。
- 编号、创建人、创建时间、更新人、更新时间为所有表的共同字段。
- 表间关系采用各表编号进行关联查询，不定义实际数据库外键。
- 本系统涉及的数据库脚本均采用 MySQL，其他数据库脚本请自行处理。

## 4.2 数据库表关系

整体数据库表间关系如图 4-1 所示。

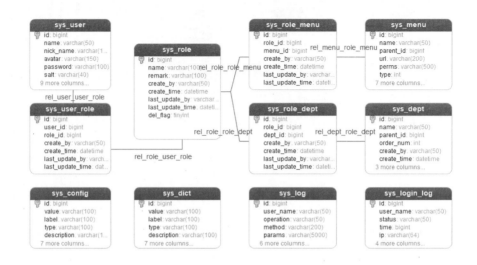

图 4-1

其中主要包含以下关系：

- 用户表和角色表通过用户角色表关联。
- 角色表和菜单表通过角色菜单表关联。
- 角色表和机构表通过角色机构表关联。
- 用户表和机构表通过用户的机构 ID 管理。
- 用户菜单和按钮权限查找流程为：用户→ 用户角色→ 角色→ 角色菜单→ 菜单。

# 4.3 数据库表结构

下面详细给出各个数据库表的建表 SQL，格式为 MySQL 数据库脚本。

## 4.3.1 用户表（sys_user）

用户表包含用户信息，主要有编号、用户名、昵称、密码、邮箱、手机号等字段，其中用户表通过表中 dept_id 与机构表关联，表明所属机构。

```
CREATE TABLE `sys_user` (
 `id` bigint(20) NOT NULL AUTO_INCREMENT COMMENT '编号',
 `name` varchar(50) NOT NULL COMMENT '用户名',
 `nick_name` varchar(150) DEFAULT NULL COMMENT '昵称',
 `avatar` varchar(150) DEFAULT NULL COMMENT '头像',
 `password` varchar(100) DEFAULT NULL COMMENT '密码',
 `salt` varchar(40) DEFAULT NULL COMMENT '加密盐',
 `email` varchar(100) DEFAULT NULL COMMENT '邮箱',
```

```
  `mobile` varchar(100) DEFAULT NULL COMMENT '手机号',
  `status` tinyint(4) DEFAULT NULL COMMENT '状态  0：禁用   1：正常',
  `dept_id` bigint(20) DEFAULT NULL COMMENT '机构ID',
  `create_by` varchar(50) DEFAULT NULL COMMENT '创建人',
  `create_time` datetime DEFAULT NULL COMMENT '创建时间',
  `last_update_by` varchar(50) DEFAULT NULL COMMENT '更新人',
  `last_update_time` datetime DEFAULT NULL COMMENT '更新时间',
  `del_flag` tinyint(4) DEFAULT '0' COMMENT '是否删除  -1：已删除  0：正常',
  PRIMARY KEY (`id`),
  UNIQUE KEY `name` (`name`)
) ENGINE=InnoDB AUTO_INCREMENT=34 DEFAULT CHARSET=utf8 COMMENT='用户管理';
```

## 4.3.2 角色表（sys_role）

角色表代表用户角色，用户拥有角色，角色拥有菜单，菜单拥有权限标识，所以不同角色拥有不同的权限，角色表主要有编号、角色名、备注等字段。

```
CREATE TABLE `sys_role` (
  `id` bigint(20) NOT NULL AUTO_INCREMENT COMMENT '编号',
  `name` varchar(100) DEFAULT NULL COMMENT '角色名称',
  `remark` varchar(100) DEFAULT NULL COMMENT '备注',
  `create_by` varchar(50) DEFAULT NULL COMMENT '创建人',
  `create_time` datetime DEFAULT NULL COMMENT '创建时间',
  `last_update_by` varchar(50) DEFAULT NULL COMMENT '更新人',
  `last_update_time` datetime DEFAULT NULL COMMENT '更新时间',
  `del_flag` tinyint(4) DEFAULT '0' COMMENT '是否删除  -1：已删除  0：正常',
  PRIMARY KEY (`id`)
) ENGINE=InnoDB AUTO_INCREMENT=9 DEFAULT CHARSET=utf8 COMMENT='角色管理';
```

## 4.3.3 机构表（sys_dept）

机构代表一种组织机构，可以有子机构，用户归属于机构。机构表主要有编号、机构名称、上级机构等字段。

```
CREATE TABLE `sys_dept` (
  `id` bigint(20) NOT NULL AUTO_INCREMENT COMMENT '编号',
  `name` varchar(50) DEFAULT NULL COMMENT '机构名称',
  `parent_id` bigint(20) DEFAULT NULL COMMENT '上级机构ID，一级机构为0',
  `order_num` int(11) DEFAULT NULL COMMENT '排序',
  `create_by` varchar(50) DEFAULT NULL COMMENT '创建人',
  `create_time` datetime DEFAULT NULL COMMENT '创建时间',
  `last_update_by` varchar(50) DEFAULT NULL COMMENT '更新人',
  `last_update_time` datetime DEFAULT NULL COMMENT '更新时间',
  `del_flag` tinyint(4) DEFAULT '0' COMMENT '是否删除  -1：已删除  0：正常',
  PRIMARY KEY (`id`)
```

```
) ENGINE=InnoDB AUTO_INCREMENT=36 DEFAULT CHARSET=utf8 COMMENT='机构管理';
```

### 4.3.4 菜单表（sys_menu）

菜单分为菜单目录、菜单和操作按钮 3 种类型，可以进行权限控制，菜单表主要有编号、菜单名称、父菜单、菜单类型、菜单图标、菜单 URL、菜单权限等字段。

```
CREATE TABLE `sys_menu` (
 `id` bigint(20) NOT NULL AUTO_INCREMENT COMMENT '编号',
 `name` varchar(50) DEFAULT NULL COMMENT '菜单名称',
 `parent_id` bigint(20) DEFAULT NULL COMMENT '父菜单ID，一级菜单为0',
 `url` varchar(200) DEFAULT NULL COMMENT '菜单URL,类型:1.普通页面（如用户管理,
/sys/user) 2.嵌套完整外部页面,以http(s)开头的链接 3.嵌套服务器页面,使用iframe:前缀+目
标URL(如SQL监控,iframe:/druid/login.html, iframe:前缀会替换成服务器地址)',
 `perms` varchar(500) DEFAULT NULL COMMENT '授权(多个用逗号分隔,如:
sys:user:add,sys:user:edit)',
 `type` int(11) DEFAULT NULL COMMENT '类型   0：目录   1：菜单   2：按钮',
 `icon` varchar(50) DEFAULT NULL COMMENT '菜单图标',
 `order_num` int(11) DEFAULT NULL COMMENT '排序',
 `create_by` varchar(50) DEFAULT NULL COMMENT '创建人',
 `create_time` datetime DEFAULT NULL COMMENT '创建时间',
 `last_update_by` varchar(50) DEFAULT NULL COMMENT '更新人',
 `last_update_time` datetime DEFAULT NULL COMMENT '更新时间',
 `del_flag` tinyint(4) DEFAULT '0' COMMENT '是否删除  -1：已删除  0：正常',
 PRIMARY KEY (`id`)
) ENGINE=InnoDB AUTO_INCREMENT=45 DEFAULT CHARSET=utf8 COMMENT='菜单管理';
```

### 4.3.5 用户角色表（sys_user_role）

用户角色表是用户和角色的中间表，通过用户 ID 和角色 ID 分别和用户表和角色表关联。

```
CREATE TABLE `sys_user_role` (
 `id` bigint(20) NOT NULL AUTO_INCREMENT COMMENT '编号',
 `user_id` bigint(20) DEFAULT NULL COMMENT '用户ID',
 `role_id` bigint(20) DEFAULT NULL COMMENT '角色ID',
 `create_by` varchar(50) DEFAULT NULL COMMENT '创建人',
 `create_time` datetime DEFAULT NULL COMMENT '创建时间',
 `last_update_by` varchar(50) DEFAULT NULL COMMENT '更新人',
 `last_update_time` datetime DEFAULT NULL COMMENT '更新时间',
 PRIMARY KEY (`id`)
) ENGINE=InnoDB AUTO_INCREMENT=76 DEFAULT CHARSET=utf8 COMMENT='用户角色';
```

### 4.3.6 角色菜单表（sys_role_menu）

角色菜单表是角色和菜单的中间表，通过角色 ID 和菜单 ID 分别和角色表和菜单表关联。

```sql
CREATE TABLE `sys_role_menu` (
  `id` bigint(20) NOT NULL AUTO_INCREMENT COMMENT '编号',
  `role_id` bigint(20) DEFAULT NULL COMMENT '角色ID',
  `menu_id` bigint(20) DEFAULT NULL COMMENT '菜单ID',
  `create_by` varchar(50) DEFAULT NULL COMMENT '创建人',
  `create_time` datetime DEFAULT NULL COMMENT '创建时间',
  `last_update_by` varchar(50) DEFAULT NULL COMMENT '更新人',
  `last_update_time` datetime DEFAULT NULL COMMENT '更新时间',
  PRIMARY KEY (`id`)
) ENGINE=InnoDB AUTO_INCREMENT=469 DEFAULT CHARSET=utf8 COMMENT='角色菜单';
```

### 4.3.7  角色机构表(sys_role_dept)

角色机构表是角色和机构的中间表,通过角色ID和机构ID分别与角色表和机构表关联。

```sql
CREATE TABLE `sys_role_dept` (
  `id` bigint(20) NOT NULL AUTO_INCREMENT COMMENT '编号',
  `role_id` bigint(20) DEFAULT NULL COMMENT '角色ID',
  `dept_id` bigint(20) DEFAULT NULL COMMENT '机构ID',
  `create_by` varchar(50) DEFAULT NULL COMMENT '创建人',
  `create_time` datetime DEFAULT NULL COMMENT '创建时间',
  `last_update_by` varchar(50) DEFAULT NULL COMMENT '更新人',
  `last_update_time` datetime DEFAULT NULL COMMENT '更新时间',
  PRIMARY KEY (`id`)
) ENGINE=InnoDB AUTO_INCREMENT=4 DEFAULT CHARSET=utf8 COMMENT='角色机构';
```

### 4.3.8  字典表(sys_dict)

字典表主要存储系统常用的枚举类型数据,主要包含编号、标签、数据值、类型等字段。

```sql
CREATE TABLE `sys_dict` (
  `id` bigint(20) NOT NULL AUTO_INCREMENT COMMENT '编号',
  `value` varchar(100) NOT NULL COMMENT '数据值',
  `label` varchar(100) NOT NULL COMMENT '标签名',
  `type` varchar(100) NOT NULL COMMENT '类型',
  `description` varchar(100) NOT NULL COMMENT '描述',
  `sort` decimal(10,0) NOT NULL COMMENT '排序(升序)',
  `create_by` varchar(50) DEFAULT NULL COMMENT '创建人',
  `create_time` datetime DEFAULT NULL COMMENT '创建时间',
  `last_update_by` varchar(50) DEFAULT NULL COMMENT '更新人',
  `last_update_time` datetime DEFAULT NULL COMMENT '更新时间',
  `remarks` varchar(255) DEFAULT NULL COMMENT '备注信息',
  `del_flag` tinyint(4) DEFAULT '0' COMMENT '是否删除  -1:已删除  0:正常',
  PRIMARY KEY (`id`)
) ENGINE=InnoDB AUTO_INCREMENT=5 DEFAULT CHARSET=utf8 COMMENT='字典表';
```

### 4.3.9 配置表（sys_config）

配置表主要存储系统配置信息，主要包含编号、标签、数据值、类型等字段。

```sql
CREATE TABLE `sys_config` (
  `id` bigint(20) NOT NULL AUTO_INCREMENT COMMENT '编号',
  `value` varchar(100) NOT NULL COMMENT '数据值',
  `label` varchar(100) NOT NULL COMMENT '标签名',
  `type` varchar(100) NOT NULL COMMENT '类型',
  `description` varchar(100) NOT NULL COMMENT '描述',
  `sort` decimal(10,0) NOT NULL COMMENT '排序（升序）',
  `create_by` varchar(50) DEFAULT NULL COMMENT '创建人',
  `create_time` datetime DEFAULT NULL COMMENT '创建时间',
  `last_update_by` varchar(50) DEFAULT NULL COMMENT '更新人',
  `last_update_time` datetime DEFAULT NULL COMMENT '更新时间',
  `remarks` varchar(255) DEFAULT NULL COMMENT '备注信息',
  `del_flag` tinyint(4) DEFAULT '0' COMMENT '是否删除  -1: 已删除   0: 正常',
  PRIMARY KEY (`id`)
) ENGINE=InnoDB AUTO_INCREMENT=5 DEFAULT CHARSET=utf8 COMMENT='系统配置表';
```

### 4.3.10 操作日志表（sys_log）

操作日志表主要记录系统用户的日常操作信息，主要包含编号、用户名、用户操作、请求方法、请求参数、执行时长、IP 地址等字段。

```sql
CREATE TABLE `sys_log` (
  `id` bigint(20) NOT NULL AUTO_INCREMENT COMMENT '编号',
  `user_name` varchar(50) DEFAULT NULL COMMENT '用户名',
  `operation` varchar(50) DEFAULT NULL COMMENT '用户操作',
  `method` varchar(200) DEFAULT NULL COMMENT '请求方法',
  `params` varchar(5000) DEFAULT NULL COMMENT '请求参数',
  `time` bigint(20) NOT NULL COMMENT '执行时长(毫秒)',
  `ip` varchar(64) DEFAULT NULL COMMENT 'IP 地址',
  `create_by` varchar(50) DEFAULT NULL COMMENT '创建人',
  `create_time` datetime DEFAULT NULL COMMENT '创建时间',
  `last_update_by` varchar(50) DEFAULT NULL COMMENT '更新人',
  `last_update_time` datetime DEFAULT NULL COMMENT '更新时间',
  PRIMARY KEY (`id`)
) ENGINE=InnoDB AUTO_INCREMENT=2798 DEFAULT CHARSET=utf8 COMMENT='系统操作日志';
```

### 4.3.11 登录日志表（sys_login_log）

登录日志表主要记录用户登录和退出状态，主要包含编号、用户名、登录状态、IP 地址等字段，可以根据 status 状态统计在线用户信息。

```sql
CREATE TABLE `sys_login_log` (
  `id` bigint(20) NOT NULL AUTO_INCREMENT COMMENT '编号',
  `user_name` varchar(50) DEFAULT NULL COMMENT '用户名',
  `status` varchar(50) DEFAULT NULL COMMENT '登录状态（online:在线，登录初始状态，方便统计在线人数；login:退出登录后将online置为login；logout:退出登录）',
  `ip` varchar(64) DEFAULT NULL COMMENT 'IP地址',
  `create_by` varchar(50) DEFAULT NULL COMMENT '创建人',
  `create_time` datetime DEFAULT NULL COMMENT '创建时间',
  `last_update_by` varchar(50) DEFAULT NULL COMMENT '更新人',
  `last_update_time` datetime DEFAULT NULL COMMENT '更新时间',
  PRIMARY KEY (`id`)
) ENGINE=InnoDB AUTO_INCREMENT=2798 DEFAULT CHARSET=utf8 COMMENT='系统登录日志';
```

# 第 5 章 搭建开发环境

## 5.1 开发环境准备

在进行项目开发前，请安装以下开发环境。

### 5.1.1 安装 JDK 环境

Java 基础开发工具和运行环境，进入后面附注的官网下载地址，根据系统下载对应的版本。我们这里选择 Windows 64 系统的 EXE 安装版本，最好选择 1.8 以上的版本，这里使用 1.8.0_131，下载安装完成之后再设置一下环境变量就可以了。网上教程很多，这里就不浪费篇幅了，下面提供官网下载地址和网上提供的安装教程。

- 下载地址：https://www.oracle.com/technetwork/java/javase/downloads/index.html
- 安装教程：https://jingyan.baidu.com/article/6dad5075d1dc40a123e36ea3.html

### 5.1.2 安装 Eclipse 开发工具

Eclipse 是一款开源的可视化开发工具，采用 OSGI 模块化架构，支持插件插拔，支持插件开发和扩展，提供非常完善和强大的功能，还有丰富和实用的插件可以选用。这里选用 Eclipse 作为后端开发 IDE，想使用 IDEA 的请自行查阅相关安装教程。下面提供官网下载地址，Eclipse 下载下来无须安装，只要有安装 Java 环境即可运行。

- 下载地址：https://www.eclipse.org/downloads/packages/

### 5.1.3 安装 MySQL 数据库

本系统采用 MySQL 数据库。MySQL 是一个非常流行的开源关系型数据库，大家也非常熟悉，网上教程也多如牛毛，这里就不再赘述了，下面提供官网下载地址和网上提供的安装教程，大家自行学习安装。

MySQL 本身不提供可视化的管理工具，但是市场上有很多免费的第三方软件可用，比如 Navicat for MySQL，读者可根据需要自行安装需要的管理工具。

- 下载地址：https://dev.mysql.com/downloads/mysql/
- 安装教程：http://www.runoob.com/mysql/mysql-install.html

### 5.1.4　安装 Maven 构建工具

本系统采用 Maven 进行项目管理和打包。Maven 是一个项目管理和综合工具。Maven 基于 POM 模型，给开发人员提供了构建一个完整的生命周期框架。开发团队可以自动完成项目的基础工具建设，Maven 使用标准的目录结构和默认构建生命周期。下面提供官网下载地址和网上提供的安装教程。

- 下载地址：http://maven.apache.org/
- 安装教程：https://www.yiibai.com/maven

## 5.2　生成项目模板

登录 Spring Initializr，输入项目信息，生成 Spring Boot 项目模板，保存到本地。网站地址为 https://start.spring.io/，页面如图 5-1 所示。

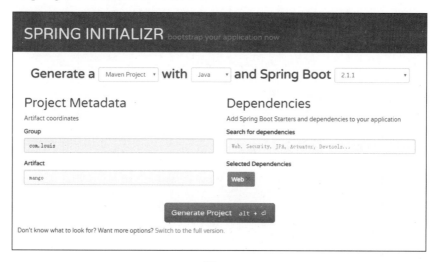

图 5-1

## 5.3　导入 Maven 项目

解压下载的 ZIP 包，放置到任意工作目录下。打开 Eclipse 开发工具，在导航栏上右击 → 导入（Import）→ Exist Maven Project，根据提示选择解压的项目目录，导入 Maven 项目到 Eclipse 开发工具，如图 5-2 所示。

图 5-2

导入成功之后,可以看到项目结构如图 5-3 所示。

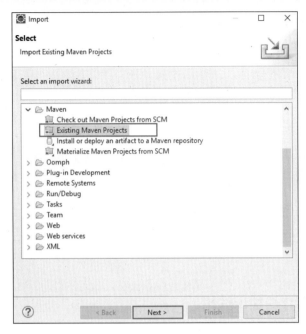

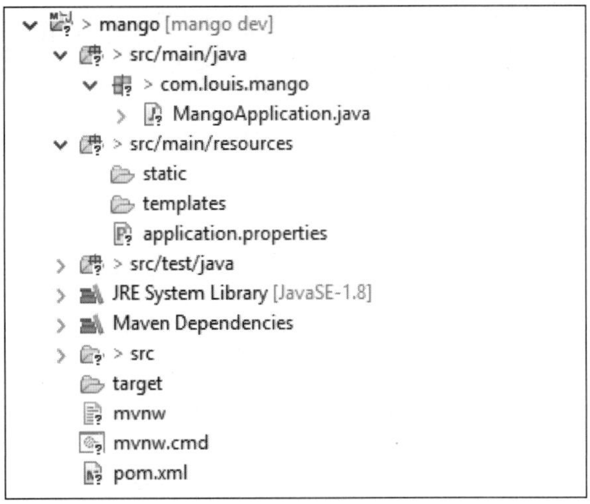

图 5-3

(1) pom.xml,Maven 的配置文件。

```xml
<?xml version="1.0" encoding="UTF-8"?>
<project xmlns="http://maven.apache.org/POM/4.0.0"
  xmlns:xsi="http://www.w3.org/2001/XMLSchema-instance"
xsi:schemaLocation="http://maven.apache.org/POM/4.0.0
http://maven.apache.org/xsd/maven-4.0.0.xsd">
    <modelVersion>4.0.0</modelVersion>
```

```xml
<groupId>com.louis</groupId>
<artifactId>mango</artifactId>
<version>0.0.1-SNAPSHOT</version>
<packaging>jar</packaging>
<name>mango</name>
<description>Demo project for Spring Boot</description>

<parent>
    <groupId>org.springframework.boot</groupId>
    <artifactId>spring-boot-starter-parent</artifactId>
    <version>2.0.4.RELEASE</version>
    <relativePath/> <!-- lookup parent from repository -->
</parent>

<properties>
    <project.build.sourceEncoding>UTF-8</project.build.sourceEncoding>
    <project.reporting.outputEncoding>UTF-8</project.reporting.outputEncoding>
    <java.version>1.8</java.version>
</properties>

<dependencies>
    <dependency>
        <groupId>org.springframework.boot</groupId>
        <artifactId>spring-boot-starter-web</artifactId>
    </dependency>
    <dependency>
        <groupId>org.springframework.boot</groupId>
        <artifactId>spring-boot-starter-test</artifactId>
        <scope>test</scope>
    </dependency>
</dependencies>

<build>
    <plugins>
        <plugin>
            <groupId>org.springframework.boot</groupId>
            <artifactId>spring-boot-maven-plugin</artifactId>
        </plugin>
    </plugins>
</build>
</project>
```

（2）application.properties，一个空的项目配置文件。

（3）MangoApplication.java，Spring Boot 应用启动类。

```
@SpringBootApplication
public class MangoApplication {

    public static void main(String[] args) {
        SpringApplication.run(MangoApplication.class, args);
    }

}
```

清理掉不需要的 mvnw、mvnw.cmd 和 test 目录下的测试文件。另外，我们这里使用 YAML 的配置格式，所以把 application.properties 改为 application.yml，清理完成后项目结构如图 5-4 所示。

图 5-4

# 5.4 编译打包运行

## 5.4.1 编译打包

右击 pom.xml，选择 run as → maven install，出现如图 5-5 所示的信息，说明编译成功。

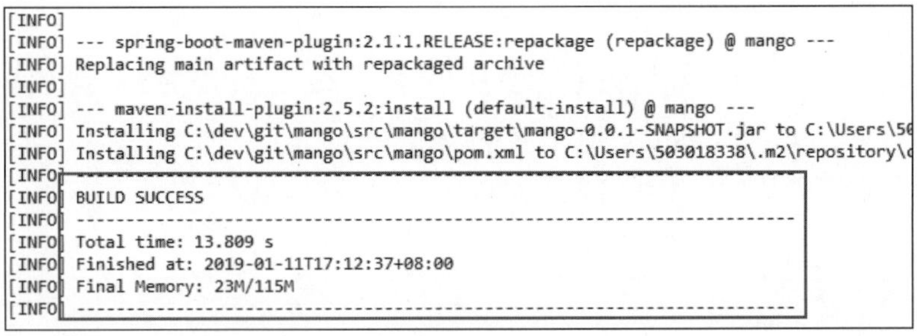

图 5-5

## 5.4.2　启动应用

右击MangoApplication.java，选择run as →Java Application，出现如图5-6所示的信息，说明启动成功。

```
  .   ____          _            __ _ _
 /\\ / ___'_ __ _ _(_)_ __  __ _ \ \ \ \
( ( )\___ | '_ | '_| | '_ \/ _` | \ \ \ \
 \\/  ___)| |_)| | | | | || (_| |  ) ) ) )
  '  |____| .__|_| |_|_| |_\__, | / / / /
 =========|_|==============|___/=/_/_/_/
 :: Spring Boot ::        (v2.1.1.RELEASE)

2019-01-11 17:07:45.174  INFO 13532 --- [           main] com.louis.mango.MangoApplication         : Starting MangoApplication on GG20J1G2E with PID 13532 (C:
2019-01-11 17:07:45.182  INFO 13532 --- [           main] com.louis.mango.MangoApplication         : No active profile set, falling back to default profiles:
2019-01-11 17:07:49.512  INFO 13532 --- [           main] o.s.b.w.embedded.tomcat.TomcatWebServer  : Tomcat initialized with port(s): 8080 (http)
2019-01-11 17:07:49.580  INFO 13532 --- [           main] o.apache.catalina.core.StandardService   : Starting service [Tomcat]
2019-01-11 17:07:49.581  INFO 13532 --- [           main] org.apache.catalina.core.StandardEngine  : Starting Servlet Engine: Apache Tomcat/9.0.13
2019-01-11 17:07:49.610  INFO 13532 --- [           main] o.a.catalina.core.AprLifecycleListener   : The APR based Apache Tomcat Native library which allows o
2019-01-11 17:07:50.096  INFO 13532 --- [           main] o.a.c.c.C.[Tomcat].[localhost].[/]       : Initializing Spring embedded WebApplicationContext
2019-01-11 17:07:50.096  INFO 13532 --- [           main] o.s.web.context.ContextLoader            : Root WebApplicationContext: initialization completed in 4
2019-01-11 17:07:50.624  INFO 13532 --- [           main] o.s.s.concurrent.ThreadPoolTaskExecutor  : Initializing ExecutorService 'applicationTaskExecutor'
2019-01-11 17:07:51.283  INFO 13532 --- [           main] o.s.b.w.embedded.tomcat.TomcatWebServer  : Tomcat started on port(s): 8080 (http) with context path
2019-01-11 17:07:51.292  INFO 13532 --- [           main] com.louis.mango.MangoApplication         : Started MangoApplication in 7.33 seconds (JVM running fo
2019-01-11 17:10:31.704  INFO 13532 --- [nio-8080-exec-1] o.a.c.c.C.[Tomcat].[localhost].[/]       : Initializing Spring DispatcherServlet 'dispatcherServlet'
2019-01-11 17:10:31.704  INFO 13532 --- [nio-8080-exec-1] o.s.web.servlet.DispatcherServlet        : Initializing Servlet 'dispatcherServlet'
2019-01-11 17:10:31.721  INFO 13532 --- [nio-8080-exec-1] o.s.web.servlet.DispatcherServlet        : Completed initialization in 16 ms
```

图 5-6

## 5.4.3　修改启动端口

可以通过修改 application.yml 中的配置属性修改应用启动端口，默认是 8080，比如这里将启动端口修改为 8001，可在配置文件中加入以下配置：

```
server:
  port: 8001
```

重新启动程序，我们看到启动端口已经换成 8001 了，如图 5-7 所示。

```
: Starting MangoApplication on GG20J1G2E with PID 24240 (C:\dev\git\mango\src`
: No active profile set, falling back to default profiles: default
: Tomcat initialized with port(s): 8001 (http)
: Starting service [Tomcat]
: Starting Servlet Engine: Apache Tomcat/9.0.13
: The APR based Apache Tomcat Native library which allows optimal performance
: Initializing Spring embedded WebApplicationContext
: Root WebApplicationContext: initialization completed in 5214 ms
: Initializing ExecutorService 'applicationTaskExecutor'
: Tomcat started on port(s): 8001 (http) with context path ''
: Started MangoApplication in 8.333 seconds (JVM running for 10.128)
```

图 5-7

## 5.4.4　自定义 Banner

Spring Boot 启动后会在控制台输出 Banner 信息，默认是显示 Spring 字样，如图 5-8 所示。

```
  .   ____          _            __ _ _
 /\\ / ___'_ __ _ _(_)_ __  __ _ \ \ \ \
( ( )\___ | '_ | '_| | '_ \/ _` | \ \ \ \
 \\/  ___)| |_)| | | | | || (_| |  ) ) ) )
  '  |____| .__|_| |_|_| |_\__, | / / / /
 =========|_|==============|___/=/_/_/_/
 :: Spring Boot ::        (v2.1.1.RELEASE)
2019-01-11 17:07:45.174  INFO 13532 --- [           main] com.louis.mango.MangoApplication
2019-01-11 17:07:45.182  INFO 13532 --- [           main] com.louis.mango.MangoApplication
2019-01-11 17:07:49.512  INFO 13532 --- [           main] o.s.b.w.embedded.tomcat.TomcatWebServer
```

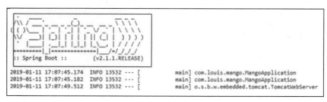

图 5-8

如果要定制自己的 Banner，只需要在 resources 下放置一个 baner.txt 文件，输入自己的 banner 字符即可，下面我们修改 mango 的 banner 信息。

Banner 字符可以通过类似以下网站生成（如图 5-9 所示）：

- http://patorjk.com/software/taag
- http://www.network-science.de/ascii/

图 5-9

复制字符到 banner.txt，并附上应用和版本信息，如图 5-10 所示。

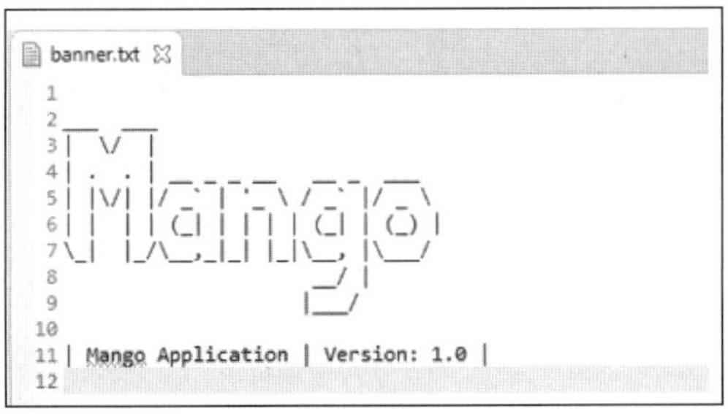

图 5-10

重新启动应用，可以看到项目的 Banner 信息已经成功替换，如图 5-11 所示。

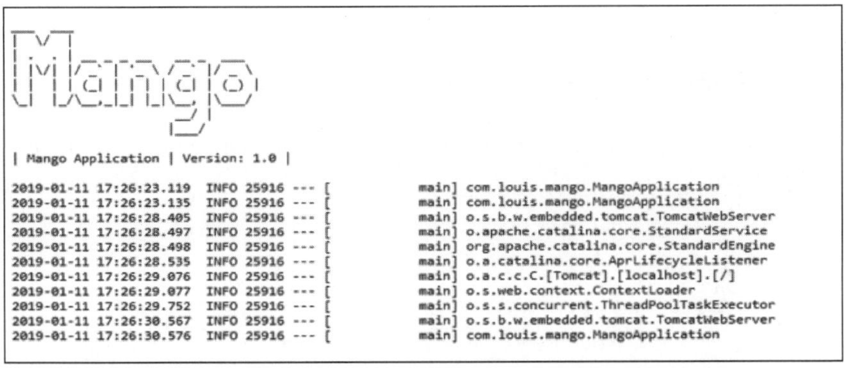

图 5-11

## 5.4.5 接口测试

启动应用，在浏览器中访问 localhost:8001，因为我们还没有提供可访问内容，所以显示没有可映射访问的内容，稍后我们添加接口进行测试，如图 5-12 所示。

图 5-12

新建一个 controller 包，并在其下创建一个 HelloController 类，添加一个 hello 接口，如图 5-13 所示。

图 5-13

```
HelloController.java
@RestController
public class HelloController {

   @GetMapping(value="/hello")
   public Object hello() {
      return "Hello Mango!";
   }

}
```

重新启动应用，在浏览器中访问 localhost:8001/hello，可以看到服务已经调用成功了，如图 5-14 所示。

图 5-14

# 第 6 章 集成Swagger文档

Spring Boot 作为当前最为流行的 Java Web 开发脚手架，越来越多的开发者选择用其来构建企业级的 RESTFul API 接口。这些接口不但会服务于传统的 Web 端（B/S），也会服务于移动端。在实际开发过程中，这些接口还要提供给开发测试进行相关的白盒测试，那么势必存在如何在多人协作中共享和及时更新 API 开发接口文档的问题。假如你已经对传统的 WIKI 文档共享方式所带来的弊端深恶痛绝，那么不妨尝试一下 Swagger2 方式，一定会让你有不一样的开发体验。

使用 Swagger 集成文档具有以下几个优势：

- 功能丰富：支持多种注解，自动生成接口文档界面，支持在界面测试 API 接口功能。
- 及时更新：开发过程中花一点写注释的时间，就可以及时地更新 API 文档，省心省力。
- 整合简单：通过添加 pom 依赖和简单配置，内嵌于应用中就可同时发布 API 接口文档界面，不需要部署独立服务。

官方网站：https://swagger.io/
官方文档：https://swagger.io/docs/

## 6.1 添加依赖

在 pom 文件内添加 Maven 依赖，这里选择 2.9.2 版本。

**pom.xml**

```xml
<!-- swagger -->
<dependency>
    <groupId>io.springfox</groupId>
    <artifactId>springfox-swagger2</artifactId>
    <version>2.9.2</version>
</dependency>
<dependency>
    <groupId>io.springfox</groupId>
    <artifactId>springfox-swagger-ui</artifactId>
```

```
    <version>2.9.2</version>
</dependency>
```

## 6.2 配置类

新建 config 包,并在其下添加 Swagger 配置类。

SwaggerConfig.java

```
@Configuration
@EnableSwagger2
public class SwaggerConfig {

    @Bean
    public Docket createRestApi(){
        return new
Docket(DocumentationType.SWAGGER_2).apiInfo(apiInfo()).select()

        .apis(RequestHandlerSelectors.any()).paths(PathSelectors.any()).build();
    }

    private ApiInfo apiInfo(){
        return new ApiInfoBuilder().build();
    }

}
```

## 6.3 页面测试

启动应用,在浏览器中访问 http://localhost:8001/swagger-ui.html#/,我们就可以看到 Swagger 的接口文档页面了,还可以选择接口进行测试,如图 6-1 所示。

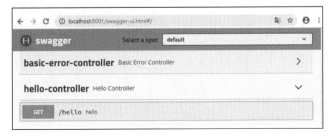

图 6-1

单击展开 hello 接口，单击右侧的 try it out→execute，发现接口成功返回 "Hello Mango！"信息，如图 6-2 所示。

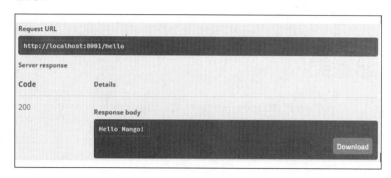

图 6-2

在后续的开发过程中，我们就可以通过 Swagger 来测试接口了。

# 第 7 章 集成MyBatis框架

MyBatis 是一款优秀的持久层框架,支持定制化 SQL、存储过程以及高级映射。MyBatis 可以使用简单的 XML 或注解来配置和映射原生信息,并将接口和 Java 的 POJOs 映射成数据库中的记录。

- 中文官网:http://www.mybatis.cn/
- 参考教程:https://www.w3cschool.cn/mybatis/

## 7.1 添加依赖

Spring Boot 对于 MyBatis 的支持需要引入 mybatis-spring-boot-starter 的依赖,修改 pom 文件,添加以下依赖。

pom.xml

```xml
<!-- mybatis -->
<dependency>
    <groupId>org.mybatis.spring.boot</groupId>
    <artifactId>mybatis-spring-boot-starter</artifactId>
    <version>1.3.2</version>
</dependency>
<!-- mysql -->
<dependency>
    <groupId>mysql</groupId>
    <artifactId>mysql-connector-java</artifactId>
</dependency>
```

## 7.2 添加配置

### 7.2.1 添加 MyBatis 配置

添加 MyBatis 配置类，配置相关扫描路径，包括 DAO、Model、XML 映射文件的扫描。在 config 包下新建一个 Mybatis 配置类。

MybatisConfig.java

```java
@Configuration
@MapperScan("com.louis.mango.**.dao")     // 扫描 DAO
public class MybatisConfig {
  @Autowired
  private DataSource dataSource;

  @Bean
  public SqlSessionFactory sqlSessionFactory() throws Exception {
    SqlSessionFactoryBean sessionFactory = new SqlSessionFactoryBean();
    sessionFactory.setDataSource(dataSource);
    sessionFactory.setTypeAliasesPackage("com.louis.mango.**.model");
                                                                // 扫描 Model

    PathMatchingResourcePatternResolver resolver = new PathMatchingResourcePatternResolver();
    sessionFactory.setMapperLocations(resolver.getResources ("classpath*:**/sqlmap/*.xml"));     // 扫描映射文件

    return sessionFactory.getObject();
  }
}
```

### 7.2.2 添加数据源配置

打开应用配置文件，添加 MySQL 数据源连接信息。

application.yml

```yaml
spring:
  datasource:
    driverClassName: com.mysql.jdbc.Driver
    url: jdbc:mysql://localhost:3306/mango?useUnicode=true&zeroDateTimeBehavior= convertToNull&autoReconnect=true&characterEncoding=utf-8
```

```
  username: root
password: 123456
```

### 7.2.3　修改启动类

给启动类 MangoApplication 的 @SpringBootApplication 注解配置包扫描，表示在应用启动时自动扫描 com.louis.mango 包下的内容，当然 Spring Boot 默认会扫描启动类包及子包的组件，所以如果启动类就是放在 com.louis.mango 下，那么默认配置其实就是 com.louis.mango 了，所以你会发现我们这里不配置包扫描，其实也是可以成功扫描到组件的。

MangoApplication.java

```
@SpringBootApplication(scanBasePackages={"com.louis.mango"})
public class MangoApplication {
    public static void main(String[] args) {
        SpringApplication.run(MangoApplication.class, args);
    }
}
```

## 7.3　生成 MyBatis 模块

手动编写 MyBatis 的 Model、DAO、XML 映射文件比较烦琐，通常都会通过一些生成工具来生成。MyBatis 官方也提供了生成工具（MyBaits Generator），另外还有一些基于官方基础改进的第三方工具，比如 MyBatis Plus 就是国内提供的一款非常优秀的开源工具，网上相关教程比较多，这里就不再赘述了。这里提供一些资料作为参考。

- Mybatis Generator 官网：http://www.mybatis.org/generator/index.html
- Mybatis Generator 教程：https://blog.csdn.net/testcs_dn/article/details/77881776
- MyBatis Plus 官网：http://mp.baomidou.com/#/
- MyBatis Plus 官网：http://mp.baomidou.com/#/quick-start

生成好代码之后，分别将 Domain、DAO、XML 映射文件复制到相应的包里，如图 7-1 所示。

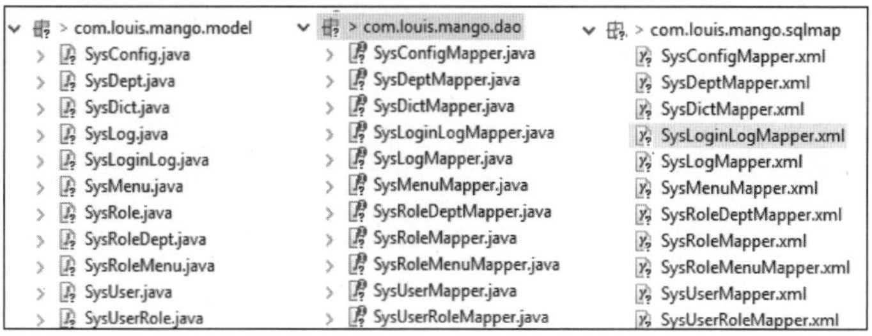

图 7-1

打开生成的 Mapper，我们可以看到 MyBatis Generator 默认生成了一些增、删、改、查的方法，我们可以直接使用。

SysUserMapper.java

```java
public interface SysUserMapper {
    int deleteByPrimaryKey(Long id);

    int insert(SysUser record);

    int insertSelective(SysUser record);

    SysUser selectByPrimaryKey(Long id);

    int updateByPrimaryKeySelective(SysUser record);

    int updateByPrimaryKey(SysUser record);
}
```

## 7.4 编写服务接口

在 Mapper 中新添加一个 findAll 方法，用于查询所有的用户信息。

SysUserMapper.java

```java
public interface SysUserMapper {
    int deleteByPrimaryKey(Long id);

    int insert(SysUser record);

    int insertSelective(SysUser record);

    SysUser selectByPrimaryKey(Long id);

    int updateByPrimaryKeySelective(SysUser record);

    int updateByPrimaryKey(SysUser record);

    /**
     * 查询全部
     * @return
     */
    List<SysUser> findAll();
}
```

在映射文件中添加一个查询方法，编写 findAll 的查询语句。

**SysUserMapper.xml**

```xml
<select id="findAll" resultMap="BaseResultMap">
  select
  <include refid="Base_Column_List" />
  from sys_user
</select>
```

然后编写用户管理接口，包含一个 findAll 方法。

**SysUserService.java**

```java
public interface SysUserService {

    /**
     * 查找所有用户
     * @return
     */
    List<SysUser> findAll();

}
```

接着编写用户管理实现类，调用 SysUserMapper 方法完成查询操作。

**SysUserServiceImpl.java**

```java
@Service
public class SysUserServiceImpl implements SysUserService {

    @Autowired
    private SysUserMapper sysUserMapper;

    @Override
    public List<SysUser> findAll() {
        return sysUserMapper.findAll();
    }
}
```

然后编写用户管理 RESTful 接口，返回 JSON 数据格式，提供外部调用。被@RestController 注解的接口控制器默认使用 JSON 格式交互，返回 JSON 结果。

**SysUserController.Java**

```java
@RestController
@RequestMapping("user")
public class SysUserController {
```

```
@Autowired
private SysUserService sysUserService;

@GetMapping(value="/findAll")
public Object findAll() {
    return sysUserService.findAll();
}

}
```

## 7.5 配置打包资源

虽然代码编写已经完成,但是此时启动运行还是会有问题的,因为在编译打包的时候,我们的 XML 映射文件是不在默认打包范围内的,所以需要修改一下打包资源配置。

修改 pom.xml,在 build 标签内加入形式如下的 resource 标签的打包配置,这样打包时就会把 MyBatis 映射文件也复制过去了。

pom.xml

```xml
<build>
    <plugins>
        <plugin>
            <groupId>org.springframework.boot</groupId>
            <artifactId>spring-boot-maven-plugin</artifactId>
        </plugin>
    </plugins>
    <!-- 打包时复制 MyBatis 的映射文件 -->
    <resources>
        <resource>
            <directory>src/main/java</directory>
            <includes>
                <include>**/sqlmap/*.xml</include>
            </includes>
            <filtering>false</filtering>
        </resource>
        <resource>
            <directory>src/main/resources</directory>
            <includes>
                <include>**/*.*</include>
            </includes>
            <filtering>true</filtering>
```

```
            </resource>
        </resources>
</build>
```

## 7.6 编译运行测试

编译并启动应用，访问 http://localhost:8001/user/findAll，可以看到查询接口成功返回了所有用户信息，如图 7-2 所示。

图 7-2

当然也可以用 Swagger 进行测试，特别是要发送 POST 请求的时候，测试起来非常方便，如图 7-3 所示。

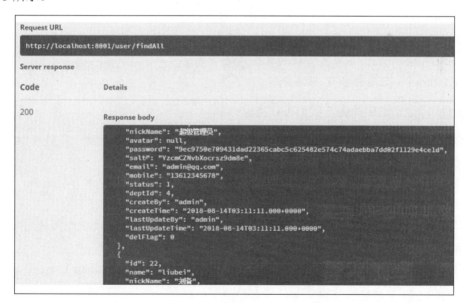

图 7-3

# 第 8 章 集成Druid数据源

数据库连接池负责分配、管理和释放数据库连接，它允许应用程序重复使用一个现有的数据库连接，而不是再重新建立一个，释放空闲时间超过最大空闲时间的数据库连接来避免因为没有释放数据库连接而引起的数据库连接遗漏。通过数据库连接池能明显提高对数据库操作的性能。在 Java 应用程序开发中，常用的连接池有 DBCP、C3P0、Proxool 等。

Spring Boot 默认提供了若干种可用的连接池，默认的数据源是 org.apache.tomcat.jdbc.pool.DataSource。Druid 是阿里系提供的一个开源连接池，除在连接池之外，还提供了非常优秀的数据库监控和扩展功能。在此，根据项目实践中的应用，讲解如何实现 Spring Boot 与 Druid 连接池的集成。

## 8.1 Druid 介绍

Druid 是阿里开源的一个 JDBC 应用组件，其主要包括 3 部分：

- DruidDriver：代理 Driver，能够提供基于 Filter-Chain 模式的插件体系。
- DruidDataSource：高效可管理的数据库连接池。
- SQLParser：实用的 SQL 语法分析。

通过 Druid 连接池中间件，我们可以实现：

- 监控数据库访问性能。Druid 内置了一个功能强大的 StatFilter 插件，能够详细统计 SQL 的执行性能，对于线上分析数据库访问性能有所帮助。
- 替换传统的 DBCP 和 C3P0 连接池中间件。Druid 提供了一个高效、功能强大、可扩展性好的数据库连接池。
- 数据库密码加密。直接把数据库密码写在配置文件中，容易导致安全问题。DruidDriver 和 DruidDataSource 都支持 PasswordCallback。
- SQL 执行日志。Druid 提供了不同的 LogFilter，能够支持 Common-Logging、Log4j 和 JdkLog，你可以按需要选择相应的 LogFilter，监控你应用的数据库访问情况。
- 扩展 JDBC。如果对 JDBC 层有编程的需求，可以通过 Druid 提供的 Filter-Chain 机制很方便地编写 JDBC 层的扩展插件。

更多详细信息可参考官方文档 https://github.com/alibaba/druid/wiki。

## 8.2 添加依赖

在 pom 文件中添加 Druid 相关的 maven 依赖。

pom.xml

```xml
<!-- druid -->
<dependency>
  <groupId>com.alibaba</groupId>
  <artifactId>druid-spring-boot-starter</artifactId>
  <version>1.1.10</version>
</dependency>
```

Druid Spring Boot Starter 是阿里官方提供的 Spring Boot 插件，用于帮助在 Spring Boot 项目中轻松集成 Druid 数据库连接池和监控。

更多资料可以参考：

- Druid：https://github.com/alibaba/druid
- Druid Spring Starter：https://github.com/alibaba/druid/tree/master/druid-spring-boot-starter

## 8.3 添加配置

修改配置文件，把原有的数据源配置替换成 Druid 数据源并配置数据源相关参数。

application.yml

```yaml
spring:
  datasource:
    name: druidDataSource
    type: com.alibaba.druid.pool.DruidDataSource
    druid:
      driver-class-name: com.mysql.jdbc.Driver
      url: jdbc:mysql://localhost:3306/mango?useUnicode=true&zeroDateTimeBehavior=convertToNull&autoReconnect=true&characterEncoding=utf-8
      username: root
      password: 123456
      filters: stat,wall,log4j,config
      max-active: 100
      initial-size: 1
```

```
            max-wait: 60000
            min-idle: 1
            time-between-eviction-runs-millis: 60000
            min-evictable-idle-time-millis: 300000
            validation-query: select 'x'
            test-while-idle: true
            test-on-borrow: false
            test-on-return: false
            pool-prepared-statements: true
            max-open-prepared-statements: 50
            max-pool-prepared-statement-per-connection-size: 20
```

参数说明：

- spring.datasource.druid.max-active：最大连接数。
- spring.datasource.druid.initial-size：初始化大小。
- spring.datasource.druid.min-idle：最小连接数。
- spring.datasource.druid.max-wait：获取连接等待超时时间。
- spring.datasource.druid.time-between-eviction-runs-millis：间隔多久才进行一次检测，检测需要关闭的空闲连接，单位是毫秒。
- spring.datasource.druid.min-evictable-idle-time-millis：一个连接在池中最小生存的时间，单位是毫秒。
- spring.datasource.druid.filters=config,stat,wall,log4j：配置监控统计拦截的 filters，去掉后监控界面 SQL 无法进行统计，wall 用于防火墙。

Druid 提供了几种 Filter 信息，如表 8-1 所示。

表 8-1 Druid 提供的几种 Filter 信息

| Filter 类名 | 别名 |
| --- | --- |
| default | com.alibaba.druid.filter.stat.StatFilter |
| stat | com.alibaba.druid.filter.stat.StatFilter |
| mergeStat | com.alibaba.druid.filter.stat.MergeStatFilter |
| encoding | com.alibaba.druid.filter.encoding.EncodingConvertFilter |
| log4j | com.alibaba.druid.filter.logging.Log4jFilter |
| log4j2 | com.alibaba.druid.filter.logging.Log4j2Filter |
| slf4j | com.alibaba.druid.filter.logging.Slf4jLogFilter |
| commonlogging | com.alibaba.druid.filter.logging.CommonsLogFilter |
| wall | com.alibaba.druid.wall.WallFilter |

如果需要通过定制的配置文件对 Druid 进行自定义属性配置，添加配置类如下：

DruidDataSourceProperties.java

```java
@ConfigurationProperties(prefix = "spring.datasource.druid")
public class DruidDataSourceProperties {
    // jdbc
    private String driverClassName;
    private String url;
    private String username;
    private String password;
    // jdbc connection pool
    private int initialSize;
    private int minIdle;
    private int maxActive = 100;
    private long maxWait;
    private long timeBetweenEvictionRunsMillis;
    private long minEvictableIdleTimeMillis;
    private String validationQuery;
    private boolean testWhileIdle;
    private boolean testOnBorrow;
    private boolean testOnReturn;
    private boolean poolPreparedStatements;
    private int maxPoolPreparedStatementPerConnectionSize;
    // filter
    private String filters;

    // 此处省略 getter 和 setter
}
```

Druid Spring Starter 简化了很多配置，如果默认配置满足不了你的需求，可以自定义配置。更多配置参考如下：

Druid Spring Starter：https://github.com/alibaba/druid/tree/master/druid-spring-boot-starter

# 8.4 配置 Servlet 和 Filter

在 config 包下新建一个 DruidConfig 配置类，主要是注入属性和配置连接池相关的配置，如黑白名单、监控管理后台登录账户密码等，内容如下：

DruidConfig.java

```java
@Configuration
@EnableConfigurationProperties({DruidDataSourceProperties.class})
```

```java
public class DruidConfig {
    @Autowired
private DruidDataSourceProperties properties;

    @Bean
    @ConditionalOnMissingBean
    public DataSource druidDataSource() {
        // 中间省略属性设置
        return druidDataSource;
    }

    /** 注册Servlet信息，配置监控视图 */
    @Bean
    @ConditionalOnMissingBean
    public ServletRegistrationBean<Servlet> druidServlet() {
        ServletRegistrationBean<Servlet> servletRegistrationBean = new
            ServletRegistrationBean<Servlet>(new StatViewServlet(),
"/druid/*");
        //白名单
        servletRegistrationBean.addInitParameter("allow",
"127.0.0.1,139.196.87.48");
        // IP黑名单 (存在共同时，deny优先于allow)：
        // 如果满足deny的话提示:Sorry, you are not permitted to view this page.
        servletRegistrationBean.addInitParameter("deny","192.168.1.119");
        //登录查看信息的账号密码,用于登录Druid监控后台
        servletRegistrationBean.addInitParameter("loginUsername", "admin");
        servletRegistrationBean.addInitParameter("loginPassword", "admin");
        //是否能够重置数据
        servletRegistrationBean.addInitParameter("resetEnable", "true");
        return servletRegistrationBean;

    }

    /** 注册Filter信息,监控拦截器 */
    @Bean
    @ConditionalOnMissingBean
    public FilterRegistrationBean<Filter> filterRegistrationBean() {
        // 省略具体注册逻辑
    }
}
```

代码说明：

- @EnableConfigurationProperties注解用于导入上一步Druid的配置信息。

- public ServletRegistrationBean druidServlet() 相当于 Web Servlet 配置。
- public FilterRegistrationBean filterRegistrationBean() 相当于 Web Filter 配置。

# 8.5 编译运行

编译启动应用，发现启动出错了，根据报错信息提示，发现缺了 log4j 的依赖，如图 8-1 所示。

```
Caused by: java.lang.NoClassDefFoundError: org/apache/log4j/Priority
        at java.lang.Class.forName0(Native Method) ~[na:1.8.0_131]
        at java.lang.Class.forName(Class.java:264) ~[na:1.8.0_131]
        at com.alibaba.druid.util.Utils.loadClass(Utils.java:203) ~[druid-1.1.10.jar:1.1.10]
        at com.alibaba.druid.filter.FilterManager.loadFilter(FilterManager.java:114) ~[druid-1.1.10.jar:1.1.10]
        at com.alibaba.druid.pool.DruidAbstractDataSource.addFilters(DruidAbstractDataSource.java:1325) ~[druid-1.1.10.jar:1.1.10]
        at com.alibaba.druid.pool.DruidAbstractDataSource.setFilters(DruidAbstractDataSource.java:1314) ~[druid-1.1.10.jar:1.1.10]
        at com.louis.mango.config.DruidConfig.druidDataSource(DruidConfig.java:54) ~[classes/:na]
        at com.louis.mango.config.DruidConfig$$EnhancerBySpringCGLIB$$f436956e.CGLIB$druidDataSource$1(<generated>) ~[classes/:na]
        at com.louis.mango.config.DruidConfig$$EnhancerBySpringCGLIB$$f436956e$$FastClassBySpringCGLIB$$b7ef3f12.invoke(<generatec
        at org.springframework.cglib.proxy.MethodProxy.invokeSuper(MethodProxy.java:244) ~[spring-core-5.1.3.RELEASE.jar:5.1.3.REL
        at org.springframework.context.annotation.ConfigurationClassEnhancer$BeanMethodInterceptor.intercept(ConfigurationClassEnh
        at com.louis.mango.config.DruidConfig$$EnhancerBySpringCGLIB$$f436956e.druidDataSource(<generated>) ~[classes/:na]
        at sun.reflect.NativeMethodAccessorImpl.invoke0(Native Method) ~[na:1.8.0_131]
        at sun.reflect.NativeMethodAccessorImpl.invoke(NativeMethodAccessorImpl.java:62) ~[na:1.8.0_131]
        at sun.reflect.DelegatingMethodAccessorImpl.invoke(DelegatingMethodAccessorImpl.java:43) ~[na:1.8.0_131]
        at java.lang.reflect.Method.invoke(Method.java:498) ~[na:1.8.0_131]
        at org.springframework.beans.factory.support.SimpleInstantiationStrategy.instantiate(SimpleInstantiationStrategy.java:154)
        ... 33 common frames omitted
Caused by: java.lang.ClassNotFoundException: org.apache.log4j.Priority
        at java.net.URLClassLoader.findClass(URLClassLoader.java:381) ~[na:1.8.0_131]
        at java.lang.ClassLoader.loadClass(ClassLoader.java:424) ~[na:1.8.0_131]
        at sun.misc.Launcher$AppClassLoader.loadClass(Launcher.java:335) ~[na:1.8.0_131]
        at java.lang.ClassLoader.loadClass(ClassLoader.java:357) ~[na:1.8.0_131]
        ... 50 common frames omitted
```

图 8-1

### 1. 添加 log4j 依赖

在 pom 文件中添加 log4j 依赖，最新版本是 1.2.17。

pom.xml

```
<!-- log4j -->
<dependency>
    <groupId>log4j</groupId>
    <artifactId>log4j</artifactId>
    <version>1.2.17</version>
</dependency>
```

### 2. 添加 log4j 配置

在 resources 目录下，新建一个 log4j 参数配置文件，并输入如下内容：

log4j.properties

```
### set log levels ###
log4j.rootLogger = INFO,DEBUG, console, infoFile, errorFile ,debugfile,mail
LocationInfo=true
```

```
log4j.appender.console = org.apache.log4j.ConsoleAppender
log4j.appender.console.Target = System.out
log4j.appender.console.layout = org.apache.log4j.PatternLayout
log4j.appender.console.layout.ConversionPattern =[%d{yyyy-MM-dd HH:mm:ss,SSS}]-[%p]:%m   %x %n

log4j.appender.infoFile = org.apache.log4j.DailyRollingFileAppender
log4j.appender.infoFile.Threshold = INFO
log4j.appender.infoFile.File = C:/logs/log
log4j.appender.infoFile.DatePattern = '.'yyyy-MM-dd'.log'
log4j.appender.infoFile.Append=true
log4j.appender.infoFile.layout = org.apache.log4j.PatternLayout
log4j.appender.infoFile.layout.ConversionPattern =[%d{yyyy-MM-dd HH:mm:ss,SSS}]-[%p]:%m   %x %n

log4j.appender.errorFile = org.apache.log4j.DailyRollingFileAppender
log4j.appender.errorFile.Threshold = ERROR
log4j.appender.errorFile.File = C:/logs/error
log4j.appender.errorFile.DatePattern = '.'yyyy-MM-dd'.log'
log4j.appender.errorFile.Append=true
log4j.appender.errorFile.layout = org.apache.log4j.PatternLayout
log4j.appender.errorFile.layout.ConversionPattern =[%d{yyyy-MM-dd HH:mm:ss,SSS}]-[%p]:%m   %x %n
```

配置完成后，重新编译启动，发现已经可以启动了。

## 8.6 查看监控

### 8.6.1 登录界面

启动应用，访问http://localhost:8001/druid/login.html，进入Druid监控后台页面，如图8-2所示。

图 8-2

## 8.6.2 监控首页

用户名就是在 DruidConfig 配置的登录账号和密码，这里的配置是"用户名：admin""密码：admin"，输入信息登录之后，Druid 后台的管理首页如图 8-3 所示。

图 8-3

## 8.6.3 数据源

数据源页显示连接数据源的相关信息，如图 8-4 所示。

图 8-4

## 8.6.4 SQL 监控

访问 http://localhost:8001/user/findAll，如图 8-5 所示。接口调用成功之后，可以看到 SQL 监控的执行记录，可以查看和分析执行的 SQL 性能，方便进行数据库性能优化。

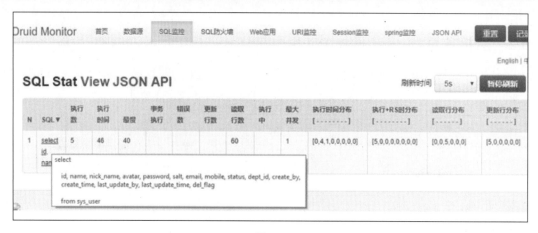

图 8-5

除此之外，Druid 监控管理后台还有 URI 监控、SQL 防火墙、Session 监控等多种多样的功能，这里就不浪费篇章进行介绍了，有兴趣的可以自行学习探索。相关链接如下：

- https://github.com/alibaba/druid/wiki
- https://github.com/alibaba/druid/tree/master/druid-spring-boot-starter

# 第 9 章 跨域解决方案

## 9.1 什么是跨域

为了保证浏览器的安全，不同源的客户端脚本在没有明确授权的情况下，不能读写对方资源。这叫作同源策略，同源策略是浏览器安全的基石。

如果一个请求地址里面的协议、域名和端口号都相同，就属于同源。

举个例子，判断下面 URL 是否和 http://www.a.com/a/a.html 同源：

- http://www.a.com/b/b.html，同源。
- http://www.b.com/a/a.html，不同源，域名不相同。
- https://www.a.com/b/b.html，不同源，协议不相同。
- http://www.a.com:8080/b/b.html，不同源，端口号不相同。

依据浏览器同源策略，非同源脚本不可操作其他源下面的对象，想要操作其他源下的对象就需要跨域。综上所述，在同源策略的限制下，非同源的网站之间不能发送 AJAX 请求。如有需要，可通过降域或其他技术实现。

## 9.2 CORS 技术

为了解决浏览器跨域问题，W3C 提出了跨源资源共享方案，即 CORS（Cross-Origin Resource Sharing）。

CORS 可以在不破坏即有规则的情况下，通过后端服务器实现 CORS 接口，从而实现跨域通信。CORS 将请求分为两类：简单请求和非简单请求，分别对跨域通信提供了支持。

### 9.2.1 简单请求

在 CORS 出现前，发送 HTTP 请求时在头信息中不能包含任何自定义字段，且 HTTP 信息不超过以下几个字段：

- Accept
- Accept-Language
- Content-Language
- Last-Event-ID
- Content-Type（仅限于[application/x-www-form-urlencoded、multipart/form-data、text/plain]类型）

一个简单请求的例子：

```
GET /test HTTP/1.1
Accept: */*
Accept-Encoding: gzip, deflate, sdch, br
Origin: http://www.test.com
Host: www.test.com
```

对于简单请求，CORS 的策略是请求时在请求头中增加一个 Origin 字段，服务器收到请求后，根据该字段判断是否允许该请求访问。

- 如果允许，就在 HTTP 头信息中添加 Access-Control-Allow-Origin 字段，并返回正确的结果。
- 如果不允许，就不在 HTTP 头信息中添加 Access-Control-Allow-Origin 字段。

除了上面提到的 Access-Control-Allow-Origin，还有几个字段用于描述 CORS 返回结果：

- Access-Control-Allow-Credentials：可选，用户是否可以发送、处理 cookie。
- Access-Control-Expose-Headers：可选，可以让用户拿到的字段。有几个字段无论设置与否都可以拿到的，包括 Cache-Control、Content-Language、Content-Type、Expires、Last-Modified、Pragma。

### 9.2.2 非简单请求

对于非简单请求的跨源请求，浏览器会在真实请求发出前增加一次 OPTION 请求，称为预检请求（preflight request）。预检请求将真实请求的信息，包括请求方法、自定义头字段、源信息添加到 HTTP 头信息字段中，询问服务器是否允许这样的操作。

例如一个 GET 请求：

```
OPTIONS /test HTTP/1.1
Origin: http://www.test.com
Access-Control-Request-Method: GET
Access-Control-Request-Headers: X-Custom-Header
Host: www.test.com
```

与 CORS 相关的字段有：

- 请求使用的 HTTP 方法 Access-Control-Request-Method。
- 请求中包含的自定义头字段 Access-Control-Request-Headers。

服务器收到请求时，需要分别对 Origin、Access-Control-Request-Method、Access-Control-Request-Headers 进行验证，验证通过后，会在返回 HTTP 头信息中添加：

```
Access-Control-Allow-Origin: http://www.test.com
Access-Control-Allow-Methods: GET, POST, PUT, DELETE
Access-Control-Allow-Headers: X-Custom-Header
Access-Control-Allow-Credentials: true
Access-Control-Max-Age: 1728000
```

它们的含义分别是：

- Access-Control-Allow-Methods：真实请求允许的方法。
- Access-Control-Allow-Headers：服务器允许使用的字段。
- Access-Control-Allow-Credentials：是否允许用户发送、处理 cookie。
- Access-Control-Max-Age：预检请求的有效期，单位为秒。有效期内，不会重复发送预检请求。

当预检请求通过后，浏览器才会发送真实请求到服务器。这样就实现了跨域资源的请求访问。

## 9.3 CORS 实现

CORS 的代码实现比较简单，主要是要理解 CORS 实现跨域的原理和方式。在 config 包下新建一个 CORS 配置类，实现 WebMvcConfigurer 接口。

### WebMvcConfigurer.java

```java
@Configuration
public class CorsConfig implements WebMvcConfigurer {

    @Override
    public void addCorsMappings(CorsRegistry registry) {
        registry.addMapping("/**")     // 允许跨域访问的路径
        .allowedOrigins("*") // 允许跨域访问的源
        .allowedMethods("POST", "GET", "PUT", "OPTIONS", "DELETE")  // 允许请求方法
        .maxAge(168000)    // 预检间隔时间
        .allowedHeaders("*")   // 允许头部设置
        .allowCredentials(true); // 是否发送 cookie
    }

}
```

这样，每当客户端发送请求的时候，都会在头部附上跨域信息，就可以支持跨域访问了。

# 第 10 章 业务功能实现

这一章节，我们开始实现各个业务功能接口，包括用户管理、机构管理、角色管理、菜单管理、字典管理、系统配置、操作日志、登录日志等业务功能的实现。

## 10.1 工程结构规划

我们要采用的是微服务的架构，虽然现在我们只有一个工程，但随着项目越来越大，代码的可重用性和可维护性就会变得越来越难，所以尽早对工程结构进行合理的规划，对项目前期的开发、后期的扩展和维护都是非常重要的。

经过重新规划，我们的工程结构如下：

- mango-common：公共代码模块，主要放置一些工具类。
- mango-core：核心业务代码模块，主要封装公共业务模块。
- mango-admin：后台管理模块，包含用户、角色、菜单管理等。
- mango-pom：聚合模块，仅为简化打包，一键执行打包所有模块。

### 10.1.1 mango-admin

将原先 mango 工程更名为 mango-admin，请遵循以下步骤进行工程重构。

（1）关闭应用，在 Eclipse 上选择删除 mango 工程，注意不要勾选删除磁盘，如图 10-1 所示。

图 10-1

（2）找到工程所在位置，修改 mango 工程名为 mango-admin，如图 10-2 所示。

图 10-2

（3）编辑 pom.xml，将需要替换的 mango 字符替换为 mango-admin，如图 10-3 所示。

图 10-3

（4）右击 Eclipse 导航栏，选择 import→exist maven project，重新导入 mango-admin 工程，如图 10-4 所示。

图 10-4

（5）重构包结构，将基础包重构成 com.louis.mango.admin，以区分不同工程的包，如图 10-5 所示。

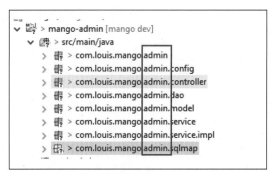

图 10-5

（6）由于重构了包路径，但是 XML 映射文件的内容无法同步更新，要将所有 MyBatis 的 XML 映射文件中出现的 Mapper 和 Model 的包路径修改为正确的路径，可以通过将 "com.louis.mango." 全部替换为 "com.louis.mango.admin" 进行统一修改，如图 10-6 所示。

```xml
<?xml version="1.0" encoding="UTF-8"?>
<!DOCTYPE mapper PUBLIC "-//mybatis.org//DTD Mapper 3.0//EN" "http://mybatis.org/dtd/mybatis
<mapper namespace="com.louis.mango.dao.SysConfigMapper">
  <resultMap id="BaseResultMap" type="com.louis.mango.model.SysConfig">
    <id column="id" jdbcType="BIGINT" property="id" />
    <result column="value" jdbcType="VARCHAR" property="value" />
    <result column="label" jdbcType="VARCHAR" property="label" />
    <result column="type" jdbcType="VARCHAR" property="type" />
    <result column="description" jdbcType="VARCHAR" property="description" />
    <result column="sort" jdbcType="DECIMAL" property="sort" />
```

图 10-6

（7）将 MangoApplication 改名为 MangoAdminApplication，编译启动应用，服务访问正常就成功了，如图 10-7 所示。

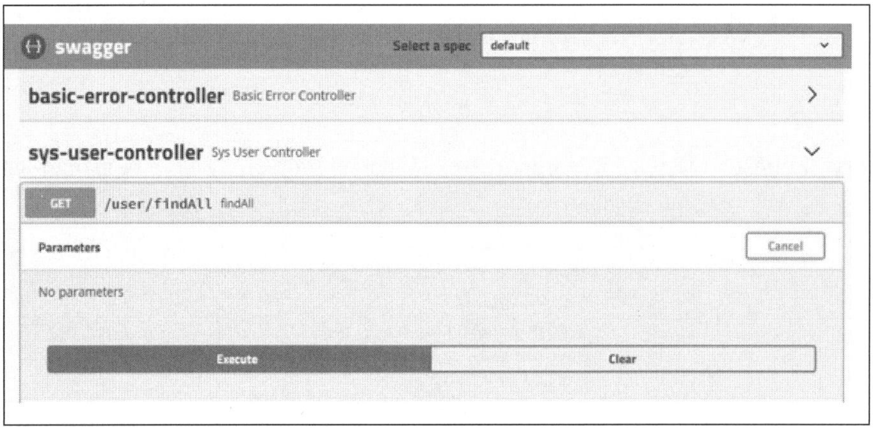

图 10-7

## 10.1.2　mango-common

新建一个空的 maven 工程，除了一个 pom.xml，没有其他内容，后续放置一些工具方法和常量，如图 10-8 所示。

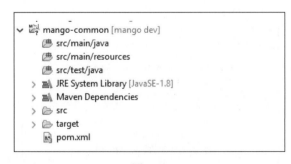

图 10-8

## 10.1.3 mango-core

新建一个空的 maven 工程，后续放置一些公共核心业务代码封装，如 HTTP 交互格式封装、业务基类封装和分页工具封装等，如图 10-9 所示。

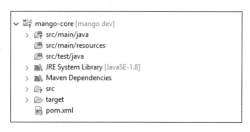

图 10-9

## 10.1.4 mango-pom

为了方便统一打包，新建一个 mango-pom 工程，如图 10-10 所示。这个工程依赖所有模块，负责统一进行打包（不然编译的时候需要逐个编译，工程一多很是麻烦），但因我们采用的是微服务架构，每个工程模块使用的依赖版本可能都是不一样的，所以这里的 mango-pom 与所有模块不存在实质性的父子模块关系，也不由 mango-pom 进行统一版本和依赖管理，只是为了便利打包。

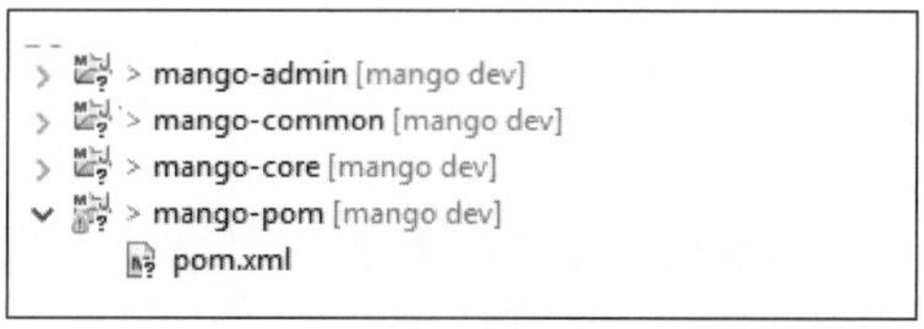

图 10-10

## 10.1.5 打包测试

下面进行统一打包测试，我们最终需要的效果是：只要在 mango-pom 下的 pom.xml 运行打包就能编译打包所有模块。现在各个子模块都没有编译过，模块间的依赖也还没有加，所以第一次还需要遵循以下步骤进行操作：

（1）右击 mango-common 下的 pom.xml，执行 maven→Run As→Maven Install 编译打包。

（2）在 mango-core 下的 pom.xml 内添加 mango-common 为 dependency 依赖，然后执行编译打包命令。

pom.xml

```
<dependency>
    <groupId>com.louis</groupId>
    <artifactId>mango-common</artifactId>
```

```xml
    <version>1.0.0</version>
</dependency>
```

（3）在 mango-admin 下的 pom.xml 内添加 mango-core 为 dependency 依赖，然后执行编译打包命令。

pom.xml

```xml
<dependency>
    <groupId>com.louis</groupId>
    <artifactId>mango-core</artifactId>
    <version>1.0.0</version>
</dependency>
```

（4）在 mango-pom 下的 pom.xml 内添加以上所有的模块为 modules 依赖，然后执行编译打包命令。

pom.xml

```xml
<modules>
    <module>../mango-admin</module>
    <module>../mango-common</module>
    <module>../mango-core</module>
</modules>
```

注意 如果工程出现红叉，可以尝试右击工程 Maven→Update Project 进行解决。

以后只要对 mango-pom 下的 pom.xml 执行命令，就可以统一打包所有模块了。如果控制台输出如下所示的打包信息，就表示打包成功了。

```
[INFO] ------------------------------------------------------------------------
[INFO] Reactor Summary:
[INFO]
[INFO] mango-common ....................................... SUCCESS [  6.765 s]
[INFO] mango-core ......................................... SUCCESS [  1.032 s]
[INFO] mango-admin ........................................ SUCCESS [ 17.707 s]
[INFO] mango-pom .......................................... SUCCESS [  0.090 s]
[INFO] ------------------------------------------------------------------------
[INFO] BUILD SUCCESS
[INFO] ------------------------------------------------------------------------
[INFO] Total time: 28.549 s
[INFO] Finished at: 2019-01-12T17:36:46+08:00
[INFO] Final Memory: 51M/256M
[INFO] ------------------------------------------------------------------------
```

## 10.2 业务代码封装

为了统一业务代码接口、保持代码整洁、提升代码性能，这里对一些通用的代码进行了统一封装，封装内容如图 10-11 所示。

图 10-11

### 10.2.1 通用 CURD 接口

CurdService 是对通用增、删、改、查接口的封装，统一定义了包含保存、删除、批量删除、根据 ID 查询和分页查询方法，一般的业务 Service 接口会继承此接口，提供基础增、删、改、查服务，这几个接口能满足大部分基础 CURD 业务的需求，封装详情参见代码注释。

**CurdService.java**

```
/**
 * 通用 CURD 接口
 * @author Louis
 * @date Jan 12, 2019
 */
public interface CurdService<T> {

    /**
     * 保存操作
     * @param record
     */
    int save(T record);

    /**
     * 删除操作
```

```
     * @param record
     */
    int delete(T record);

    /**
     * 批量删除操作
     * @param entities
     */
    int delete(List<T> records);

    /**
     * 根据ID查询
     * @param id
     */
    T findById(Long id);

    /**
         * 分页查询
     * 这里统一封装了分页请求和结果,避免直接引入具体框架的分页对象,
     * 如MyBatis或JPA的分页对象从而避免因为替换ORM框架而导致服务层、
     * 控制层的分页接口也需要变动的情况,替换ORM框架也不会影响服务层
     * 以上的分页接口,起到了解耦的作用
     * @param pageRequest 自定义,统一分页查询请求
     * @return PageResult 自定义,统一分页查询结果
         */
    PageResult findPage(PageRequest pageRequest);
}
```

## 10.2.2 分页请求封装

对分页请求的参数进行了统一封装,传入分页查询的页码和数量即可。

PageRequest.java

```
/**
 * 分页请求
 * @author Louis
 * @date Jan 12, 2019
 */
public class PageRequest {
    /**
     * 当前页码
     */
    private int pageNum = 1;
```

```java
/**
 * 每页数量
 */
private int pageSize = 10;
/**
 * 查询参数
 */
private Map<String, Object> params = new HashMap<>();

// 此处省略 getter 和 setter
}
```

### 10.2.3 分页结果封装

对分页查询的结果进行了统一封装，结果返回业务数据和分页数据。

PageResult.java

```java
/**
 * 分页返回结果
 * @author Louis
 * @date Jan 12, 2019
 */
public class PageResult {
    /**
     * 当前页码
     */
    private int pageNum;
    /**
     * 每页数量
     */
    private int pageSize;
    /**
     * 记录总数
     */
    private long totalSize;
    /**
     * 页码总数
     */
    private int totalPages;
    /**
     * 分页数据
     */
    private List<?> content;
```

```
    // 此处省略 getter 和 setter
}
```

### 10.2.4　分页助手封装

对 MyBatis 的分页查询业务代码进行了统一封装,通过分页助手可以极大简化 Service 查询业务的编写。

MybatisPageHelper.java

```java
/**
 * MyBatis 分页查询助手
 * @author Louis
 * @date Jan 12, 2019
 */
public class MybatisPageHelper {

    public static final String findPage = "findPage";

    /**
     * 分页查询,约定查询方法名为 "findPage"
     * @param pageRequest 分页请求
     * @param mapper Dao 对象,MyBatis 的 Mapper
     * @param args 方法参数
     */
    public static PageResult findPage(PageRequest pageRequest, Object mapper) {
        return findPage(pageRequest, mapper, findPage);
    }

    /**
     * 调用分页插件进行分页查询
     * @param pageRequest 分页请求
     * @param mapper Dao 对象,MyBatis 的 Mapper
     * @param queryMethodName 要分页的查询方法名
     * @param args 方法参数
     */
    @SuppressWarnings({ "unchecked", "rawtypes" })
    public static PageResult findPage(PageRequest pageRequest, Object mapper, String queryMethodName, Object... args) {
        // 设置分页参数
        int pageNum = pageRequest.getPageNum();
        int pageSize = pageRequest.getPageSize();
        PageHelper.startPage(pageNum, pageSize);
```

```
        // 利用反射调用查询方法
        Object result = ReflectionUtils.invoke(mapper, queryMethodName, args);
        return getPageResult(pageRequest, new PageInfo((List) result));
    }

    /**
     * 将分页信息封装到统一的接口
     * @param pageRequest
     * @param page
     */
    private static PageResult getPageResult(PageRequest pageRequest,
PageInfo<?> pageInfo) {
        PageResult pageResult = new PageResult();
          pageResult.setPageNum(pageInfo.getPageNum());
          pageResult.setPageSize(pageInfo.getPageSize());
        pageResult.setTotalSize(pageInfo.getTotal());
        pageResult.setTotalPages(pageInfo.getPages());
        pageResult.setContent(pageInfo.getList());
        return pageResult;
    }
}
```

## 10.2.5 HTTP 结果封装

对接口调用的返回结果进行了统一封装，方便前端或移动端对返回结果进行统一处理。

HttpResult.java

```
/**
 * HTTP 结果封装
 * @author Louis
 * @date Jan 12, 2019
 */
public class HttpResult {

    private int code = 200;
    private String msg;
    private Object data;

    public static HttpResult error() {
        return error(HttpStatus.SC_INTERNAL_SERVER_ERROR,
            "未知异常，请联系管理员");
    }

    public static HttpResult error(String msg) {
        return error(HttpStatus.SC_INTERNAL_SERVER_ERROR, msg);
```

```java
    }
    public static HttpResult error(int code, String msg) {
        HttpResult r = new HttpResult();
        r.setCode(code);
        r.setMsg(msg);
        return r;
    }
    public static HttpResult ok(String msg) {
        HttpResult r = new HttpResult();
        r.setMsg(msg);
        return r;
    }
    public static HttpResult ok(Object data) {
        HttpResult r = new HttpResult();
        r.setData(data);
        return r;
    }
    public static HttpResult ok() {
        return new HttpResult();
    }
    // 此处省略getter和setter
}
```

## 10.3 MyBatis 分页查询

使用 MyBatis 时，最头痛的就是写分页，需要先写一个查询 count 的 select 语句，再写一个真正分页查询的语句，当查询条件多了之后，就会发现真不想花双倍的时间写 count 和 select。幸好我们有 pagehelper 分页插件。pagehelper 是一个强大实用的 MyBatis 分页插件，可以帮助我们快速地实现分页功能。那么，接下来我们就一起来体验一下吧。

### 10.3.1 添加依赖

在 mango-core 下的 pom.xml 文件内添加分页插件依赖包，因为 mango-admin 模块依赖 mango-core 模块，所以 mango-admin 模块也能获取分页插件依赖，这就是 Maven 管理依赖的好处。

pom.xml

```xml
<!-- pagehelper -->
<dependency>
    <groupId>com.github.pagehelper</groupId>
    <artifactId>pagehelper-spring-boot-starter</artifactId>
    <version>1.2.5</version>
</dependency>
```

### 10.3.2 添加配置

在 mango-admin 配置文件内添加分页插件配置。

application.yml

```yaml
# pagehelper
pagehelper:
  helperDialect: mysql
  reasonable: true
  supportMethodsArguments: true
params: count=countSql
```

### 10.3.3 分页代码

首先，我们在 DAO 层添加一个分页查询方法。

SysUserMapper.java

```java
/**
 * 分页查询
 * @return
 */
List<SysUser> findPage();
```

给 SysUserMapper.xml 添加查询方法，这是一个普通的查找全部记录的查询语句，并不需要写分页 SQL，分页插件会拦截查询请求，并读取前台传来的分页查询参数重新生成分页查询语句。

SysUserMapper.xml

```xml
<select id="findPage" resultMap="BaseResultMap">
  select
  <include refid="Base_Column_List" />
  from sys_user
</select>
```

服务层调用 DAO 层完成分页查询，让 SysUserService 继承 CurdService 接口。

### SysUserService.java

```java
public interface SysUserService extends CurdService<SysUser> {
    /**
     * 查找所有用户
     * @return
     */
    List<SysUser> findAll();
}
```

在实现类中编写分页查询业务实现，我们可以看到，经过对分页查询业务的封装，普通分页查询非常简单，只需调用 MybatisPageHelper.findPage(pageRequest, sysUserMapper)一行代码即可完成分页查询功能。

### SysUserServiceImpl.java

```java
@Service
public class SysUserServiceImpl implements SysUserService {
    @Autowired
    private SysUserMapper sysUserMapper;

    @Override
    public PageResult findPage(PageRequest pageRequest) {
        return MybatisPageHelper.findPage(pageRequest, sysUserMapper);
    }

    // 因为我们关注的是分页查询，此处省略其他方式实现
}
```

编写分页查询接口，简单调用 Service 的查询接口。

### SysUserController.java

```java
@RestController
@RequestMapping("user")
public class SysUserController {

    @Autowired
    private SysUserService sysUserService;

    @PostMapping(value="/findPage")
    public HttpResult findPage(@RequestBody PageRequest pageRequest) {
        return HttpResult.ok(sysUserService.findPage(pageRequest));
    }
}
```

## 10.3.4　接口测试

重新编译启动应用，访问 http://localhost:8001/swagger-ui.html，进入 Swagger 测试页，如图 10-12 所示。

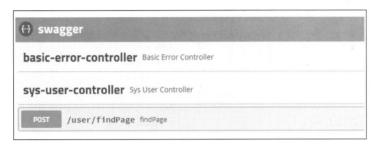

图 10-12

分别输入不同的分页查询参数，查看返回的分页结果，如果没有问题就可以了，如图 10-13 所示。

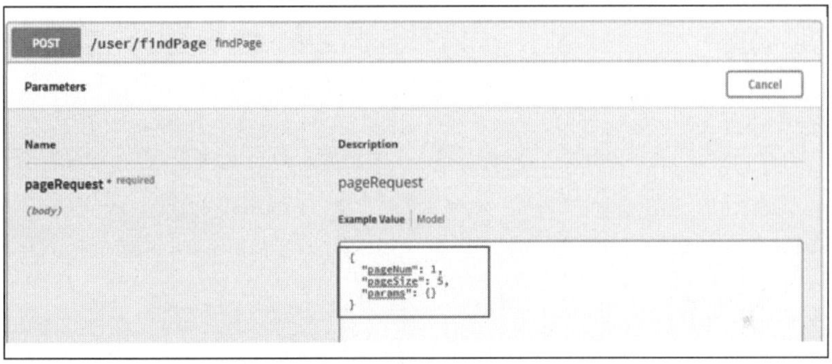

图 10-13

返回结果示例，查询参数 { pageNum: 1, pageSize: 3 }，如图 10-14 所示。

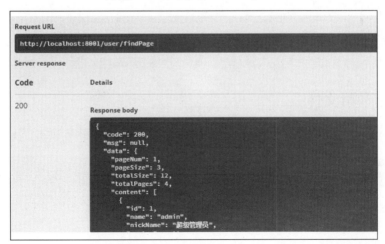

图 10-14

# 10.4 业务功能开发

业务功能的开发基本都是各种增、删、改、查业务的编写，大多都是重复性工作，业务编写也没有多少技术要点可讲，这里就拿机构管理的开发来作为基础 CURD 的开发范例。

首先，需要事先规划一下，根据需求设计好需要的接口，比如字典管理除了通用的保存、删除、分页查询接口外，还需要一个根据标签名称查询记录的查询方法。

## 10.4.1 编写 DAO 接口

打开 DAO 接口，添加 findPage、findByLable 和 findPageByLabel 三个接口。

SysDictMapper.java

```java
/**
 * 分页查询
 * @param label
 */
List<SysDict> findPage();

/**
 * 根据标签名称查询
 * @param label
 */
List<SysDict> findByLable(@Param(value="label") String label);

/**
 * 根据标签名称分页查询
 * @param label
 */
List<SysDict> findPageByLabel(@Param(value="label") String label);
```

## 10.4.2 编写映射文件

打开映射文件，编写 3 个查询方法：findPage、findByLable 和 findPageByLabel。

SysDictMappe.xml

```xml
<select id="findPage" resultMap="BaseResultMap">
  select
  <include refid="Base_Column_List" />
  from sys_dict
</select>
```

```xml
<select id="findPageByLabel" parameterType="java.lang.String"
resultMap="BaseResultMap">
    <bind name="pattern" value="'%' + _parameter.label + '%'" />
    select
    <include refid="Base_Column_List" />
    from sys_dict
    where label like #{pattern}
</select>
<select id="findByLable" parameterType="java.lang.String"
resultMap="BaseResultMap">
    select
    <include refid="Base_Column_List" />
    from sys_dict
    where label = #{label,jdbcType=VARCHAR}
</select>
```

### 10.4.3 编写服务接口

新建一个字典接口并继承通用业务接口 CurdService,额外添加一个 findByLable 接口。

SysDictService.java

```java
/**
 * 字典管理
 * @author Louis
 * @date Jan 13, 2019
 */
public interface SysDictService extends CurdService<SysDict> {

    /**
     * 根据名称查询
     * @param lable
     * @return
     */
    List<SysDict> findByLable(String lable);
}
```

### 10.4.4 编写服务实现

新建一个实现类并实现 SysDictService,调用 DAO 实现相应的业务功能。

SysDictServiceImpl.java

```java
@Service
public class SysDictServiceImpl implements SysDictService {
```

```java
    @Autowired
    private SysDictMapper sysDictMapper;

    @Override
    public int save(SysDict record) {
        if(record.getId() == null || record.getId() == 0) {
            return sysDictMapper.insertSelective(record);
        }
        return sysDictMapper.updateByPrimaryKeySelective(record);
    }

    @Override
    public int delete(SysDict record) {
        return sysDictMapper.deleteByPrimaryKey(record.getId());
    }

    @Override
    public int delete(List<SysDict> records) {
        for(SysDict record:records) {
            delete(record);
        }
        return 1;
    }

    @Override
    public SysDict findById(Long id) {
        return sysDictMapper.selectByPrimaryKey(id);
    }

    @Override
    public PageResult findPage(PageRequest pageRequest) {
        Object label = pageRequest.getParam("label");
        if(label!= null) {
            return MybatisPageHelper.findPage(pageRequest, sysDictMapper, "findPageByLabel", label);
        }
        return MybatisPageHelper.findPage(pageRequest, sysDictMapper);
    }

    @Override
    public List<SysDict> findByLable(String lable) {
        return sysDictMapper.findByLable(lable);
    }
}
```

## 10.4.5 编写控制器

新建一个字段管理控制器，注入 Service 并调用 Service 的方法实现业务接口。

SysDictController.java

```java
@RestController
@RequestMapping("dict")
public class SysDictController {

    @Autowired
    private SysDictService sysDictService;

    @PostMapping(value="/save")
    public HttpResult save(@RequestBody SysDict record) {
        return HttpResult.ok(sysDictService.save(record));
    }

    @PostMapping(value="/delete")
    public HttpResult delete(@RequestBody List<SysDict> records) {
        return HttpResult.ok(sysDictService.delete(records));
    }

    @PostMapping(value="/findPage")
    public HttpResult findPage(@RequestBody PageRequest pageRequest) {
        return HttpResult.ok(sysDictService.findPage(pageRequest));
    }

    @GetMapping(value="/findByLable")
    public HttpResult findByLable(@RequestParam String lable) {
        return HttpResult.ok(sysDictService.findByLable(lable));
    }
}
```

其他业务功能还有诸如用户管理、角色管理、机构管理、菜单管理、系统日志等业务都与此同理，这里就不再浪费篇幅了，读者根据自身需求有针对性地查阅相关代码即可。

## 10.5 业务接口汇总

### 10.5.1 用户管理

#### 1. 保存用户

接口 URL：/user/save
接口参数：SysRole
返回结果：HttpResult
接口描述：保存记录

#### 2. 删除用户

接口 URL：/user/delete
接口参数：SysRole 集合
返回结果：HttpResult
接口描述：删除记录

#### 3. 分页查询

接口 URL：/user/findPage
接口参数：PageRequest
返回结果：HttpResult
接口描述：分页查询

#### 4. 根据名称查询

接口 URL：/user/findByName
接口参数：String name
返回结果：HttpResult
接口描述：根据名称查询

#### 5. 查询用户权限

接口 URL：/user/findPermissions
接口参数：String name
返回结果：HttpResult
接口描述：查询用户权限

6. 查询用户角色

接口 URL：/user/findUserRoles
接口参数：Long userId
返回结果：HttpResult
接口描述：查询用户权限

### 10.5.2 机构管理

1. 保存机构

接口 URL：/dept/save
接口参数：SysDept
返回结果：HttpResult
接口描述：保存记录

2. 删除机构

接口 URL：/dept/delete
接口参数：SysDept 集合
返回结果：HttpResult
接口描述：删除记录

3. 查询机构树

接口 URL：/dept/findTree
接口参数：无
返回结果：HttpResult
接口描述：查询机构树

### 10.5.3 角色管理

1. 保存角色

接口 URL：/role/save
接口参数：SysRole
返回结果：HttpResult
接口描述：保存记录

2. 删除角色

接口 URL：/role/delete
接口参数：SysRole 集合
返回结果：HttpResult

接口描述：删除记录

### 3. 分页查询

接口URL：/role/findPage
接口参数：PageRequest
返回结果：HttpResult
接口描述：分页查询

### 4. 查询全部

接口URL：/role/findAll
接口参数：PageRequest
返回结果：HttpResult
接口描述：查询全部角色

### 5. 查询角色菜单

接口URL：/role/ findRoleMenus
接口参数：Long roleId
返回结果：HttpResult
接口描述：查询角色菜单

### 6. 保存角色菜单

接口URL：/role/ saveRoleMenus
接口参数：SysRoleMenu 集合
返回结果：HttpResult
接口描述：保存角色菜单

## 10.5.4 菜单管理

### 1. 保存菜单

接口URL：/menu/save
接口参数：SysMenu
返回结果：HttpResult
接口描述：保存记录

### 2. 删除菜单

接口URL：/menu/delete
接口参数：SysMenu 集合
返回结果：HttpResult
接口描述：删除记录

### 3. 查询导航菜单树

接口 URL：/menu/findNavTree
接口参数：String userName
返回结果：HttpResult
接口描述：查询导航菜单树

### 4. 查询菜单树

接口 URL：/menu/findMenuTree
接口参数：无
返回结果：HttpResult
接口描述：查询菜单树

## 10.5.5 字典管理

### 1. 保存字典

接口 URL：/dict/save
接口参数：SysDict
返回结果：HttpResult
接口描述：保存记录

### 2. 删除字典

接口 URL：/dict/delete
接口参数：SysDict 集合
返回结果：HttpResult
接口描述：删除记录

### 3. 分页查询

接口 URL：/dict/findPage
接口参数：PageRequest
返回结果：HttpResult
接口描述：分页查询

### 4. 根据标签查询

接口 URL：/dict/ findByLable
接口参数：String lable
返回结果：HttpResult
接口描述：根据标签名称查询

### 10.5.6 系统配置

#### 1. 保存配置

接口 URL：/config/save
接口参数：SysConfig
返回结果：HttpResult
接口描述：保存记录

#### 2. 删除字典

接口 URL：/config/delete
接口参数：SysConfig 集合
返回结果：HttpResult
接口描述：删除记录

#### 3. 分页查询

接口 URL：/config/findPage
接口参数：PageRequest
返回结果：HttpResult
接口描述：分页查询

#### 4. 根据标签查询

接口 URL：/config/ findByLable
接口参数：String lable
返回结果：HttpResult
接口描述：根据标签名称查询

### 10.5.7 登录日志

#### 1. 分页查询

接口 URL：/loginlog/findPage
接口参数：PageRequest
返回结果：HttpResult
接口描述：分页查询

#### 2. 删除操作日志

接口 URL：/ loginlog/delete
接口参数：List<SysLoginLog>

返回结果：HttpResult

接口描述：清除登录日志

### 10.5.8 操作日志

#### 1. 分页查询

接口 URL：/ log/findPage

接口参数：PageRequest

返回结果：HttpResult

接口描述：分页查询

#### 2. 删除操作日志

接口 URL：/ log/delete

接口参数：List<SysLog>

返回结果：HttpResult

接口描述：清除操作日志

## 10.6 导出 Excel 报表

在实际项目中，报表导出是非常普遍的需求，特别是 Excel 报表，对数据的汇总和传递都非常便利，Apache POI 是 Apache 软件基金会的开放源码函式库，POI 提供 API 给 Java 程序对 Microsoft Office 格式档案读和写的功能。这里，我们将使用 POI 实现用户信息的 Excel 报表作为范例进行讲解，后续读者有其他的报表导出需求可参考此范例。

- 官网地址：http://poi.apache.org/
- 相关教程：https://www.yiibai.com/apache_poi/

### 10.6.1 添加依赖

在 mango-common 下的 pom 文件中添加 POI 的相关依赖包，最新版本是 4.0.1。

pom.xml

```xml
<!-- poi -->
<dependency>
    <groupId>org.apache.poi</groupId>
    <artifactId>poi-ooxml</artifactId>
    <version>4.0.1</version>
</dependency>
```

## 10.6.2 编写服务接口

在用户管理接口中添加一个导出用户信息 Excel 报表的方法，采用分页查询的方式，可以传入要导出数据的范围，如需要导出全部，把页数据调至很大即可，同时因为调用的是分页查询方法查询数据，所以同样支持传入过滤字段进行数据过滤。

SysUserService.java

```java
/**
 * 生成用户信息 Excel 文件
 * @param pageRequest 要导出的分页查询参数
 * @return
 */
File createUserExcelFile (PageRequest pageRequest);
```

## 10.6.3 编写服务实现

在用户管理服务实现类中编写实现代码，生成 Excel 文件。

SysUserServiceImpl.java

```java
@Override
public File createUserExcelFile(PageRequest pageRequest) {
    PageResult pageResult = findPage(pageRequest);
    return createUserExcelFile(pageResult.getContent());
}

public static File createUserExcelFile(List<?> records) {
    if (records == null) {
        records = new ArrayList<>();
    }
    Workbook workbook = new XSSFWorkbook();
    Sheet sheet = workbook.createSheet();
    Row row0 = sheet.createRow(0);
    int columnIndex = 0;
    row0.createCell(columnIndex).setCellValue("No");
    row0.createCell(++columnIndex).setCellValue("ID");
    row0.createCell(++columnIndex).setCellValue("用户名");
    row0.createCell(++columnIndex).setCellValue("昵称");
    row0.createCell(++columnIndex).setCellValue("机构");
    row0.createCell(++columnIndex).setCellValue("角色");
    row0.createCell(++columnIndex).setCellValue("邮箱");
    row0.createCell(++columnIndex).setCellValue("手机号");
    row0.createCell(++columnIndex).setCellValue("状态");
```

```
        row0.createCell(++columnIndex).setCellValue("头像");
        row0.createCell(++columnIndex).setCellValue("创建人");
        row0.createCell(++columnIndex).setCellValue("创建时间");
        row0.createCell(++columnIndex).setCellValue("最后更新人");
        row0.createCell(++columnIndex).setCellValue("最后更新时间");
        for (int i = 0; i < records.size(); i++) {
            SysUser user = (SysUser) records.get(i);
            Row row = sheet.createRow(i + 1);
            for (int j = 0; j < columnIndex + 1; j++) {
                row.createCell(j);
            }
            columnIndex = 0;
            row.getCell(columnIndex).setCellValue(i + 1);
            row.getCell(++columnIndex).setCellValue(user.getId());
            row.getCell(++columnIndex).setCellValue(user.getName());
            row.getCell(++columnIndex).setCellValue(user.getNickName());
            row.getCell(++columnIndex).setCellValue(user.getDeptName());
            row.getCell(++columnIndex).setCellValue(user.getRoleNames());
            row.getCell(++columnIndex).setCellValue(user.getEmail());
            row.getCell(++columnIndex).setCellValue(user.getStatus());
            row.getCell(++columnIndex).setCellValue(user.getAvatar());
            row.getCell(++columnIndex).setCellValue(user.getCreateBy());

    row.getCell(++columnIndex).setCellValue(DateTimeUtils.getDateTime(user.get
CreateTime()));
            row.getCell(++columnIndex).setCellValue(user.getLastUpdateBy());

    row.getCell(++columnIndex).setCellValue(DateTimeUtils.getDateTime(user.get
LastUpdateTime()));
        }
        return PoiUtils.createExcelFile(workbook, "download_user");
}
```

### 10.6.4 编写控制器

在用户管理控制器类中添加一个接口,并调用 Service 获取 File,最终通过文件操作工具类将 File 下载到本地。

#### SysUserController.java

```
@PostMapping(value="/exportExcelUser")
public void exportExcelUser(@RequestBody PageRequest pageRequest,
HttpServletResponse res) {
    File file = sysUserService.createUserExcelFile(pageRequest);
```

```
        FileUtils.downloadFile(res, file, file.getName());
}
```

### 10.6.5　工具类代码

为了简化代码，前面代码的实现封装了一些工具类。为了便于理解，这里把关键的代码贴出来。

#### 1. PoiUtils

在编写服务实现的时候我们通过 PoiUtils 中的 createExcelFile 方法生成 Excel 文件。

```
/**
 * POI 相关操作
 * @author Louis
 * @date Jan 14, 2019
 */
public class PoiUtils {

    /**
     * 生成Excel文件
     * @param workbook
     * @param fileName
     */
    public static File createExcelFile(Workbook workbook, String fileName) {
        OutputStream stream = null;
        File file = null;
        try {
            file = File.createTempFile(fileName, ".xlsx");
            stream = new FileOutputStream(file.getAbsoluteFile());
            workbook.write(stream);
        } catch (FileNotFoundException e) {
            e.printStackTrace();
        } catch (IOException e) {
            e.printStackTrace();
        } finally {
            IOUtils.closeQuietly(workbook);
            IOUtils.closeQuietly(stream);
        }
        return file;
    }
}
```

#### 2. FileUtils

在编写导出接口的时候我们通过 FileUtils 中的 downloadFile 将 Excel 文件下载到本地。

```java
/**
 * 文件相关操作
 * @author Louis
 * @date Jan 14, 2019
 */
public class FileUtils {

    /**
     * 下载文件
     * @param response
     * @param file
     * @param newFileName
     */
    public static void downloadFile(HttpServletResponse response, File file,
String newFileName) {
        try {
            response.setHeader("Content-Disposition", "attachment; filename="
                    + new String(newFileName.getBytes("ISO-8859-1"), "UTF-8"));
            BufferedOutputStream bos =
                    new BufferedOutputStream(response.getOutputStream());
            InputStream is = new FileInputStream(file.getAbsolutePath());
            BufferedInputStream bis = new BufferedInputStream(is);
            int length = 0;
            byte[] temp = new byte[1 * 1024 * 10];
            while ((length = bis.read(temp)) != -1) {
                bos.write(temp, 0, length);
            }
            bos.flush();
            bis.close();
            bos.close();
            is.close();
        } catch (Exception e) {
            e.printStackTrace();
        }
    }
}
```

### 10.6.6　接口测试

编译启动应用，访问 http://localhost:8001/swagger-ui.html，进入 Swagger 接口测试页。

输入分页查询信息，指定要导出的用户数据范围，单击 Execute 按钮发送请求，如图 10-15 所示。

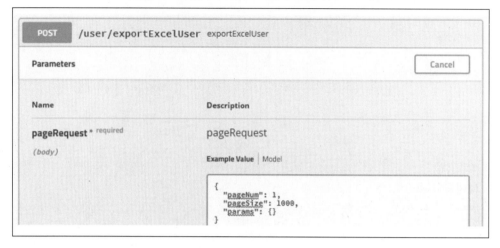

图 10-15

最终成功导出的用户报表文件内容如图 10-16 所示。

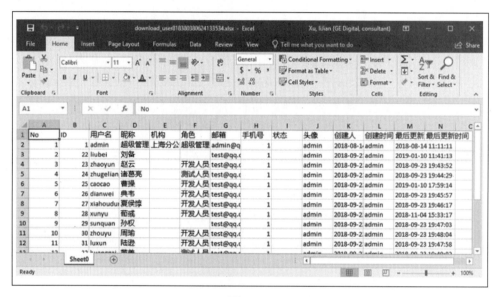

图 10-16

# 第 11 章 登录流程实现

用户登录流程是后台管理系统必备的功能，接下来我们将实现用户登录流程。在这个过程中，我们还将利用 kaptcha 实现登录验证码，利用 Spring Security 进行安全控制。

## 11.1 登录验证码

登录辅助验证是大多数登录系统都会用到的一个功能，验证方式也是多种多样的，比如登录验证码、登录验证条以及拼图拖动块等，我们这里就针对输入登录验证码的方式来实现一个范例。kaptcha 是一个开源的验证码实现库，利用这个库可以非常方便地实现验证码功能。

### 11.1.1 添加依赖

在 mango-admin 下的 pom 文件添加 kaptcha 依赖包，最新版本是 0.0.9。

pom.xml

```xml
<!-- kaptcha -->
<dependency>
    <groupId>com.github.axet</groupId>
    <artifactId>kaptcha</artifactId>
    <version>.0.9</version>
</dependency>
```

### 11.1.2 添加配置

在 config 包下创建一个 kaptcha 配置类，配置验证码的一些生成属性。

KaptchaConfig.java

```java
/**
 * 验证码配置
 * @author Louis
 * @date Jan 14, 2019
 */
@Configuration
```

```java
public class KaptchaConfig {

    @Bean
    public DefaultKaptcha producer() {
        Properties properties = new Properties();
        properties.put("kaptcha.border", "no");
        properties.put("kaptcha.textproducer.font.color", "black");
        properties.put("kaptcha.textproducer.char.space", "5");
        Config config = new Config(properties);
        DefaultKaptcha defaultKaptcha = new DefaultKaptcha();
        defaultKaptcha.setConfig(config);
        return defaultKaptcha;
    }
}
```

### 11.1.3 生成代码

新建一个控制器,提供系统登录相关的 API,在其中添加验证码生成接口。

SysLoginController.java

```java
@RestController
public class SysLoginController {
    @Autowired
    private Producer producer;

    @GetMapping("captcha.jpg")
    public void captcha(HttpServletResponse response, HttpServletRequest request) throws ServletException, IOException {
        response.setHeader("Cache-Control", "no-store, no-cache");
        response.setContentType("image/jpeg");
        // 生成文字验证码
        String text = producer.createText();
        // 生成图片验证码
        BufferedImage image = producer.createImage(text);
        // 保存到验证码到 session
        request.getSession().setAttribute(Constants.KAPTCHA_SESSION_KEY, text);
        ServletOutputStream out = response.getOutputStream();
        ImageIO.write(image, "jpg", out);
        IOUtils.closeQuietly(out);
    }
}
```

## 11.1.4 接口测试

编译启动应用，访问 http://localhost:8001/swagger-ui.html，进入 Swagger 测试页，如图 11-1 所示。

图 11-1

# 11.2 Spring Security

在 Web 应用开发中，安全一直是非常重要的一个方面。Spring Security 基于 Spring 框架，提供了一套 Web 应用安全性的完整解决方案。JWT（JSON Web Token）是当前比较主流的 Token 令牌生成方案，非常适合作为登录和授权认证的凭证。这里我们就使用 Spring Security 并结合 JWT 实现用户认证（Authentication）和用户授权（Authorization）两个主要部分的安全内容。

- JWT 官网：https://jwt.io/introduction/
- Spring Security 官网：https://spring.io/projects/spring-security
- Spring Security 教程：https://www.w3cschool.cn/springsecurity/

## 11.2.1 添加依赖

在 mango-admin 下的 pom 文件中添加 Spring Security 和 JWT 依赖包，jwt 目前的最新版本是 0.9.1。

pox.ml

```
<!-- spring security -->
<dependency>
    <groupId>org.springframework.boot</groupId>
    <artifactId>spring-boot-starter-security</artifactId>
</dependency>
<!-- jwt -->
<dependency>
```

```xml
        <groupId>io.jsonwebtoken</groupId>
        <artifactId>jjwt</artifactId>
        <version>0.9.1</version>
</dependency>
```

### 11.2.2 添加配置

在 config 包下新建一个 Spring Security 的配置类 WebSecurityConfig，主要是进行一些安全相关的配置，比如权限 URL 匹配策略、认证过滤器配置、定制身份验证组件、开启权限认证注解等，具体代码作用参见代码注释。

WebSecurityConfig.java

```java
@Configuration
@EnableWebSecurity // 开启 Spring Security
@EnableGlobalMethodSecurity(prePostEnabled = true) // 开启权限注解，如：
@PreAuthorize 注解
public class WebSecurityConfig extends WebSecurityConfigurerAdapter {

    @Autowired
    private UserDetailsService userDetailsService;

    @Override
    public void configure(AuthenticationManagerBuilder auth) throws Exception {
        // 使用自定义身份验证组件
        auth.authenticationProvider(new JwtAuthenticationProvider(userDetailsService));
    }

    @Override
    protected void configure(HttpSecurity http) throws Exception {
        // 禁用 csrf, 由于使用的是 JWT，我们这里不需要 csrf
        http.cors().and().csrf().disable().authorizeRequests()
        // 跨域预检请求
            .antMatchers(HttpMethod.OPTIONS, "/**").permitAll()
            // web jars
            .antMatchers("/webjars/**").permitAll()
            // 查看 SQL 监控（druid）
            .antMatchers("/druid/**").permitAll()
            // 首页和登录页面
            .antMatchers("/").permitAll().antMatchers("/login").permitAll()
            // swagger
            .antMatchers("/swagger-ui.html").permitAll()
            .antMatchers("/swagger-resources/**").permitAll()
```

```
            .antMatchers("/v2/api-docs").permitAll()
            .antMatchers("/webjars/springfox-swagger-ui/**").permitAll()
            // 验证码
            .antMatchers("/captcha.jpg**").permitAll()
            // 服务监控
            .antMatchers("/actuator/**").permitAll()
            // 其他所有请求需要身份认证
            .anyRequest().authenticated();
        // 退出登录处理器
        http.logout().logoutSuccessHandler(new
HttpStatusReturningLogoutSuccessHandler());
        // token 验证过滤器
        http.addFilterBefore(new
JwtAuthenticationFilter(authenticationManager()),
UsernamePasswordAuthenticationFilter.class);
    }

    @Bean
    @Override
    public AuthenticationManager authenticationManager() throws Exception {
        return super.authenticationManager();
    }
}
```

## 11.2.3 登录认证过滤器

登录认证过滤器负责登录认证时检查并生产令牌保存到上下文,接口权限认证过程时,系统从上下文获取令牌校验接口访问权限,新建一个 security 包,在其下创建 JwtAuthenticationFilter 并继承 BasicAuthenticationFilter,覆写其中的 doFilterInternal 方法进行 Token 校验。

JwtAuthenticationFilter.java

```
/**
 * 登录认证过滤器
 * @author Louis
 * @date Jan 14, 2019
 */
public class JwtAuthenticationFilter extends BasicAuthenticationFilter {

    @Autowired
    public JwtAuthenticationFilter(AuthenticationManager authenticationManager) {
        super(authenticationManager);
    }
```

```java
    @Override
protected void doFilterInternal(HttpServletRequest request, HttpServletResponse
response, FilterChain chain) throws IOException, ServletException {
    // 获取token, 并检查登录状态
        SecurityUtils.checkAuthentication(request);
        chain.doFilter(request, response);
    }
}
```

这里我们把验证逻辑抽取到了 SecurityUtils 的 checkAuthentication 方法中，checkAuthentication 通过 JwtTokenUtils 的方法获取认证信息并保存到 Spring Security 上下文。

SecurityUtils.java

```java
/**
 * 获取令牌进行认证
 * @param request
 */
public static void checkAuthentication(HttpServletRequest request) {
    // 获取令牌并根据令牌获取登录认证信息
    Authentication authentication = JwtTokenUtils.getAuthenticationeFromToken(request);
    // 设置登录认证信息到上下文
    SecurityContextHolder.getContext().setAuthentication(authentication);
}
```

JwtTokenUtils 的 getAuthenticationeFromToken 方法获取并校验请求携带的令牌。

JwtTokenUtils.java

```java
/**
 * 根据请求令牌获取登录认证信息
 * @param token 令牌
 * @return 用户名
 */
public static Authentication getAuthenticationeFromToken(HttpServletRequest request) {
    Authentication authentication = null;
    // 获取请求携带的令牌
    String token = JwtTokenUtils.getToken(request);
    if(token != null) {
        // 请求令牌不能为空
        if(SecurityUtils.getAuthentication() == null) {
            // 上下文中 Authentication 为空
            Claims claims = getClaimsFromToken(token);
```

```java
            if(claims == null) {
                return null;
            }
            String username = claims.getSubject();
            if(username == null) {
                return null;
            }
            if(isTokenExpired(token)) {
                return null;
            }
            Object authors = claims.get(AUTHORITIES);
            List<GrantedAuthority> authorities = new ArrayList<GrantedAuthority>();
            if (authors != null && authors instanceof List) {
                for (Object object : (List) authors) {
                    authorities.add(new GrantedAuthorityImpl(
                            (String) ((Map) object).get("authority")));
                }
            }
            authentication = new JwtAuthenticatioToken(username, null,
                    authorities, token);
        } else {
            if(validateToken(token, SecurityUtils.getUsername())) {
                // 如果上下文中Authentication非空,且请求令牌合法,
                // 直接返回当前登录认证信息
                authentication = SecurityUtils.getAuthentication();
            }
        }
    }
    return authentication;
}
```

JwtTokenUtils 的 getToken 尝试从请求头中获取请求携带的令牌,默认从请求头中的"Authorization"参数以"Bearer "开头的信息为令牌信息,若为空则尝试从"token"参数获取。

JwtTokenUtils.java

```java
/**
 * 获取请求token
 * @param request
 * @return
 */
public static String getToken(HttpServletRequest request) {
    String token = request.getHeader("Authorization");
    String tokenHead = "Bearer ";
```

```
    if(token == null) {
        token = request.getHeader("token");
    } else if(token.contains(tokenHead)){
        token = token.substring(tokenHead.length());
    }
    if("".equals(token)) {
        token = null;
    }
    return token;
}
```

### 11.2.4 身份验证组件

Spring Security 的登录验证是交由 ProviderManager 负责的，ProviderManager 在实际验证的时候又会通过调用 AuthenticationProvider 的 authenticate 方法来进行认证。数据库类型的默认实现方案是 DaoAuthenticationProvider。我们这里通过继承 DaoAuthenticationProvider 定制默认的登录认证逻辑，在 Security 包下新建验证器 JwtAuthenticationProvider 并继承 DaoAuthenticationProvider，覆盖实现 additionalAuthenticationChecks 方法进行密码匹配，我们这里没有使用默认的密码认证器（我们使用盐 salt 来对密码加密，默认密码验证器没有加盐），所以在这里定制了自己的密码校验逻辑，当然你也可以通过直接覆写 authenticate 方法来完成更大范围的登录认证需求定制。

JwtAuthenticationProvider.java

```java
/**
 * 身份验证提供者
 * @author Louis
 * @date Jan 14, 2019
 */
public class JwtAuthenticationProvider extends DaoAuthenticationProvider {

    public JwtAuthenticationProvider(UserDetailsService userDetailsService) {
        setUserDetailsService(userDetailsService);
    }

    @Override
    protected void additionalAuthenticationChecks(UserDetails userDetails,
UsernamePasswordAuthenticationToken authentication)
    throws AuthenticationException {
        if (authentication.getCredentials() == null) {
            logger.debug("Authentication failed: no credentials provided");
            throw new BadCredentialsException(
                messages.getMessage("AbstractUserDetailsAuthenticationProvider
```

```
                .badCredentials", "Bad credentials"));
        }
        String presentedPassword = authentication.getCredentials().toString();
        String salt = ((JwtUserDetails) userDetails).getSalt();
        if (!new PasswordEncoder(salt).matches(userDetails.getPassword(),
                presentedPassword)) {    // 覆写密码验证逻辑
            logger.debug("Authentication failed: password does not match");
            throw new BadCredentialsException(messages.getMessage(
                "AbstractUserDetailsAuthenticationProvider.badCredentials",
                "Bad credentials"));
        }
    }
}
```

### 11.2.5　认证信息查询

我们上面提到登录认证默认是通过 DaoAuthenticationProvider 来完成登录认证的,而我们知道登录验证器在进行时肯定是要从数据库获取用户信息进行匹配的,而这个获取用户信息的任务是通过 Spring Security 的 UserDetailsService 组件来完成的。

在 security 包下新建一个 UserDetailsServiceImpl 并实现 UserDetailsService 接口,覆写其中的方法 loadUserByUsername,查询用户的密码信息和权限信息并封装到 UserDetails 的实现类对象,作为结果 JwtUserDetails 返回给 DaoAuthenticationProvider 做进一步处理。

UserDetailsServiceImpl.java

```
/**
 * 用户登录认证信息查询
 * @author Louis
 * @date Jan 14, 2019
 */
@Service
public class UserDetailsServiceImpl implements UserDetailsService {

    @Autowired
    private SysUserService sysUserService;

    @Override
    public UserDetails loadUserByUsername(String username) throws
UsernameNotFoundException {
        SysUser user = sysUserService.findByName(username);
        if (user == null) {
            throw new UsernameNotFoundException("该用户不存在");
        }
        // 用户权限列表,根据权限标识如@PreAuthorize("hasAuthority('sys:menu:view')")
```

```java
        // 标注的接口对比，决定是否可以调用接口
        Set<String> permissions = sysUserService.findPermissions(user.getName());
        List<GrantedAuthority> grantedAuthorities = permissions.stream()
            .map(GrantedAuthorityImpl::new).collect(Collectors.toList());
        return new JwtUserDetails(user.getName(), user.getPassword(), user.getSalt(),
            grantedAuthorities);
    }
}
```

JwtUserDetails 是对认证信息的封装，实现 Spring Security，提供 UserDetails 接口，主要包含用户名、密码、加密盐和权限信息。

JwtUserDetails.java

```java
/**
 * 安全用户模型
 * @author Louis
 * @date Jan 14, 2019
 */
public class JwtUserDetails implements UserDetails {
    private String username;
    private String password;
    private String salt;
    private Collection<? extends GrantedAuthority> authorities;

    // 此处省略其他信息
}
```

GrantedAuthorityImpl 实现 Spring Security 的 GrantedAuthority 接口，是对权限的封装，内部包含一个字符串类型的权限标识 authority，对应菜单表的 perms 字段的权限字符串，比如用户管理的增、删、改、查权限标志 sys:user:view、sys:user:add、sys:user:edit、sys:user:delete。

GrantedAuthorityImpl.java

```java
/**
 * 权限封装
 * @author Louis
 * @date Jan 14, 2019
 */
public class GrantedAuthorityImpl implements GrantedAuthority {

    private String authority;

    public GrantedAuthorityImpl(String authority) {
        this.authority = authority;
```

```
    }

    public void setAuthority(String authority) {
        this.authority = authority;
    }

    @Override
    public String getAuthority() {
        return this.authority;
    }
}
```

## 11.2.6 添加权限注解

在用户拥有某个后台接口访问权限的时候才能访问,这叫作接口保护。我们这里就通过 Spring Security 提供的权限注解来保护后台接口免受非法访问,这里以字典管理模块为例,其他模块同理。

在 SysDictController 的接口上添加类似@PreAuthorize("hasAuthority('sys:dict:view')")的注解,表示只有当前登录用户拥有'sys:dict:view'权限标识才能访问此接口,这里的权限标识需对应菜单表中的 perms 权限信息,所以可以通过配置菜单表的权限来灵活控制接口的访问权限。

**SysDictController.java**

```
@RestController
@RequestMapping("dict")
public class SysDictController {

    @Autowired
    private SysDictService sysDictService;

    @PreAuthorize("hasAuthority('sys:dict:add') AND hasAuthority('sys:dict:edit')")
    @PostMapping(value="/save")
    public HttpResult save(@RequestBody SysDict record) {
        return HttpResult.ok(sysDictService.save(record));
    }

    @PreAuthorize("hasAuthority('sys:dict:delete')")
    @PostMapping(value="/delete")
    public HttpResult delete(@RequestBody List<SysDict> records) {
        return HttpResult.ok(sysDictService.delete(records));
    }

    @PreAuthorize("hasAuthority('sys:dict:view')")
```

```java
    @PostMapping(value="/findPage")
    public HttpResult findPage(@RequestBody PageRequest pageRequest) {
        return HttpResult.ok(sysDictService.findPage(pageRequest));
    }

    @PreAuthorize("hasAuthority('sys:dict:view')")
    @GetMapping(value="/findByLable")
    public HttpResult findByLable(@RequestParam String lable) {
        return HttpResult.ok(sysDictService.findByLable(lable));
    }
}
```

### 11.2.7　Swagger 添加令牌参数

由于我们引入 Spring Security 安全框架，接口受到保护，需要携带合法的 token 令牌（一般是登录成功之后由后台返回）才能正常访问，但是 Swagger 本身的接口测试页面默认并没有提供传送 token 参数的地方，因此需要简单定制一下，修改 SwaggerConfig 配置类即可。

SwaggerConfig.java

```java
/**
 * Swagger 配置
 * @author Louis
 * @date Jan 11, 2019
 */
@Configuration
@EnableSwagger2
public class SwaggerConfig {

    @Bean
    public Docket createRestApi(){
        // 添加请求参数，我们这里把 token 作为请求头部参数传入后端
        ParameterBuilder parameterBuilder = new ParameterBuilder();
        List<Parameter> parameters = new ArrayList<Parameter>();
        parameterBuilder.name("token").description("令牌")
        .modelRef(new ModelRef("string")).parameterType("header").
required(false).build();
        parameters.add(parameterBuilder.build());
        return new Docket(DocumentationType.SWAGGER_2).apiInfo(apiInfo()).select()
        .apis(RequestHandlerSelectors.any()).paths(PathSelectors.any())
        .build().globalOperationParameters(parameters);
    }

    private ApiInfo apiInfo(){
```

```
        return new ApiInfoBuilder().build();
    }
}
```

配置之后，重新启动就可以进行传参了，测试接口时先将登录接口返回的 token 复制到此处即可，如图 11-2 所示。Swagger 在发送请求的时候会把 token 放入请求头，后续篇幅还会讲到 token 相关的内容。

图 11-2

## 11.3 登录接口实现

在登录控制器中添加一个登录接口 login，在其中验证验证码、用户名、密码信息。匹配成功之后，执行 Spring Security 的登录认证机制。登录成功之后，返回 Token 令牌凭证。

SysLoginController.java

```
/**
 * 登录接口
 */
@PostMapping(value = "/login")
public HttpResult login(@RequestBody LoginBean loginBean, HttpServletRequest request) throws IOException {
    String username = loginBean.getAccount();
    String password = loginBean.getPassword();
    String captcha = loginBean.getCaptcha();
    // 从 session 中获取之前保存的验证码，跟前台传来的验证码进行匹配
    Object kaptcha = request.getSession().getAttribute(Constants.KAPTCHA_SESSION_KEY);
    if(kaptcha == null){
```

```
        return HttpResult.error("验证码已失效");
    }
    if(!captcha.equals(kaptcha)){
        return HttpResult.error("验证码不正确");
    }
    // 用户信息
    SysUser user = sysUserService.findByName(username);
    // 账号不存在、密码错误
    if (user == null) {
        return HttpResult.error("账号不存在");
    }
    if (!PasswordUtils.matches(user.getSalt(), password, user.getPassword())) {
        return HttpResult.error("密码不正确");
    }
    // 账号锁定
    if (user.getStatus() == 0) {
        return HttpResult.error("账号已被锁定,请联系管理员");
    }
    // 系统登录认证
    JwtAuthenticatioToken token = SecurityUtils.login(request, username, password, authenticationManager);
    return HttpResult.ok(token);
}
```

我们这里将 Spring Security 的登录认证逻辑封装到了工具类 SecurityUtils 的 login 方法中，认证流程大致分为下面 4 个步骤：

（1）将用户名密码的认证信息封装到 JwtAuthenticatioToken 对象。
（2）通过调用 authenticationManager.authenticate(token)执行认证流程。
（3）通过 SecurityContextHolder 将认证信息保存到 Security 上下文。
（4）通过 JwtTokenUtils.generateToken(authentication)生成 token 并返回。

具体过程详见代码。

**SecurityUtils.java**

```
/**
 * 系统登录认证
 * @param request
 * @param username
 * @param password
 * @param authenticationManager
 */
public static JwtAuthenticatioToken login(HttpServletRequest request, String username, String password, AuthenticationManager authenticationManager) {
```

```
    JwtAuthenticatioToken token = new JwtAuthenticatioToken(username, password);
    token.setDetails(new WebAuthenticationDetailsSource().buildDetails(request));
    // 执行登录认证过程
    Authentication authentication = authenticationManager.authenticate(token);
    // 认证成功,存储认证信息到上下文
    SecurityContextHolder.getContext().setAuthentication(authentication);
    // 生成令牌并返回给客户端
    token.setToken(JwtTokenUtils.generateToken(authentication));
    return token;
}
```

关于 JwtTokenUtils 如何生成 token 的逻辑参见下面两个方法。

JwtTokenUtils.java

```
/**
 * 生成令牌
 * @param userDetails 用户
 * @return 令牌
 */
public static String generateToken(Authentication authentication) {
    Map<String, Object> claims = new HashMap<>(3);
    claims.put(USERNAME, SecurityUtils.getUsername(authentication));
    claims.put(CREATED, new Date());
    claims.put(AUTHORITIES, authentication.getAuthorities());
    return generateToken(claims);
}

/**
 * 从数据声明生成令牌
 * @param claims 数据声明
 * @return 令牌
 */
private static String generateToken(Map<String, Object> claims) {
    Date expirationDate = new Date(System.currentTimeMillis() + EXPIRE_TIME);
    return Jwts.builder().setClaims(claims).setExpiration(expirationDate)
.signWith(SignatureAlgorithm.HS512, SECRET).compact();
}
```

LoginBean 是对登录认证信息的简单封装,包含账号密码和验证码信息。

LoginBean.java

```java
/**
 * 登录接口封装对象
 * @author Louis
 * @date Oct 29, 2018
 */
public class LoginBean {
    private String account;
    private String password;
    private String captcha;

    // 此处省略 getter 和 setter
}
```

JwtAuthenticatioToken 继承 UsernamePasswordAuthenticationToken，是对令牌信息的简单封装，用来作为认证和授权的信任凭证，其中的 token 信息由 JWT 负责生成。

JwtAuthenticatioToken.java

```java
/**
 * 自定义令牌对象
 * @author Louis
 * @date Jan 14, 2019
 */
public class JwtAuthenticatioToken extends UsernamePasswordAuthenticationToken {

    private static final long serialVersionUID = 1L;

    private String token;

    public JwtAuthenticatioToken(Object principal, Object credentials){
        super(principal, credentials);
    }

    public JwtAuthenticatioToken(Object principal, Object credentials, String token){
        super(principal, credentials);
        this.token = token;
    }

    public JwtAuthenticatioToken(Object principal, Object credentials, Collection<? extends GrantedAuthority> authorities, String token) {
        super(principal, credentials, authorities);
        this.token = token;
```

```
    }

    // 此处省略 setter 和 getter
}
```

## 11.4 接口测试

编译启动应用，访问 http://localhost:8001/swagger-ui.html，进入 Swagger 测试页面。

我们先在未登录的情况下找一个接口来测试一下，检测接口权限保护有没有生效。

我们选择角色管理的 findAll 接口进行测试，因为不用传参比较方便。发送请求结果返回 403 错误，提示 "Access Denied" 信息，如图 11-3 所示。如你所想，没错，这个就是 Spring Security 针对缺少访问权限的错误反馈信息，说明我们的接口已经受到安全框架的保护。

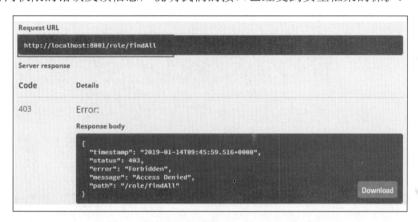

图 11-3

接着，我们访问一下验证码的生成接口，记住登录验证码，然后进行登录测试，如图 11-4 所示。

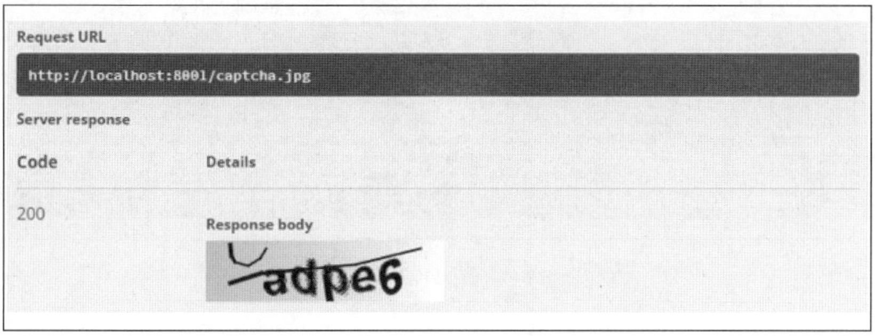

图 11-4

找到登录接口，输入用户名和密码，先输入一个错误的验证码，结果如图 11-5 所示。
然后我们输入用户名验证码和一个错误的密码，测试一下，结果如图 11-6 所示。

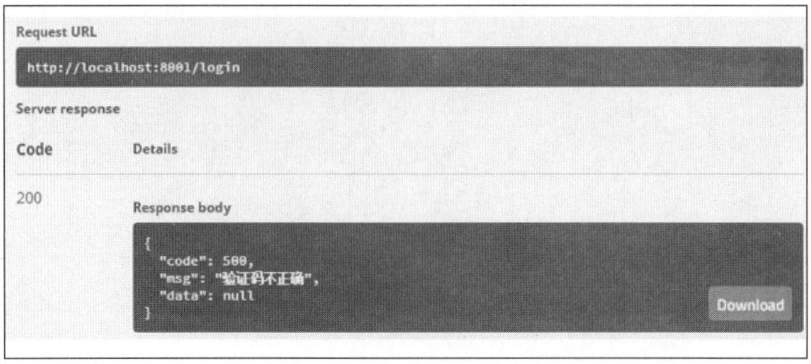

图 11-5

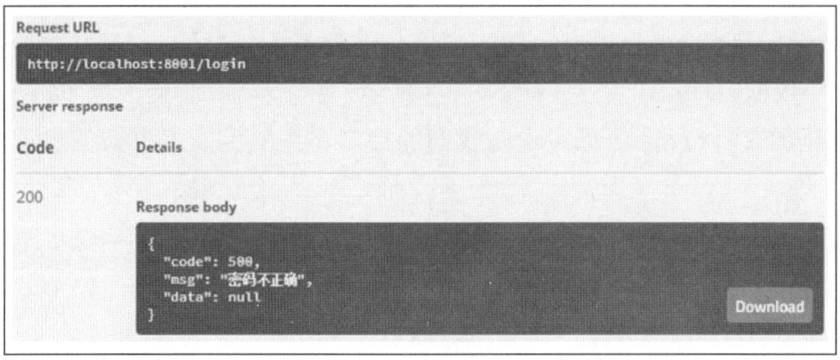

图 11-6

最后我们输入正确的信息,即"用户名:admin""密码:admin"及上一步生成的验证码,结果如图 11-7 所示。

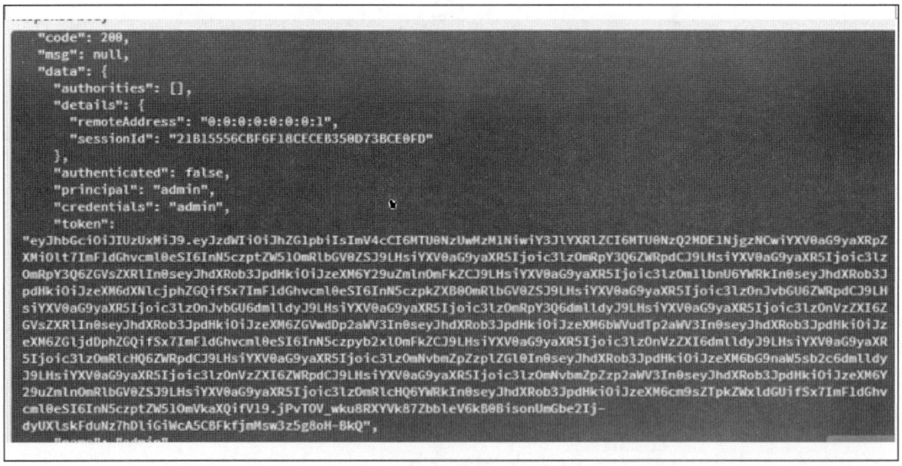

图 11-7

可以看到登录成功之后返回的登录认证信息,其中包含 JWT 生成的 token 令牌。后续访问需要携带此 token 作为凭证进行接口访问。

复制此处生成的 token 信息("token"属性的长字符串),再次找到角色管理的 findAll 接口,把复制的 token 信息粘贴到 token 令牌参数输入框内,如图 11-8 所示。

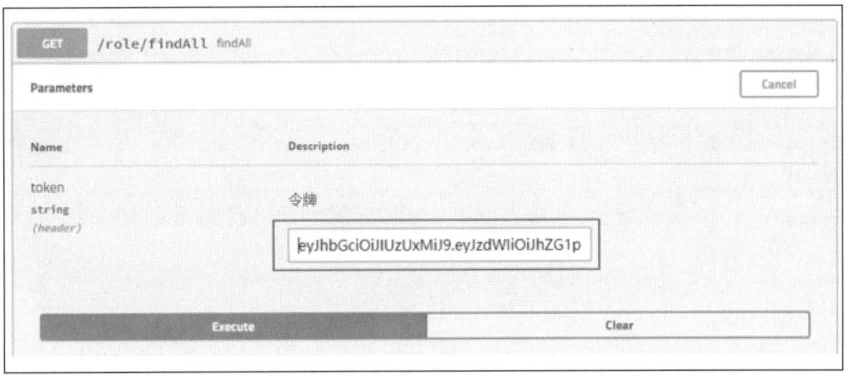

图 11-8

单击 Execute 按钮发送查询请求，发现成功地返回了角色数据，结果如图 11-9 所示。

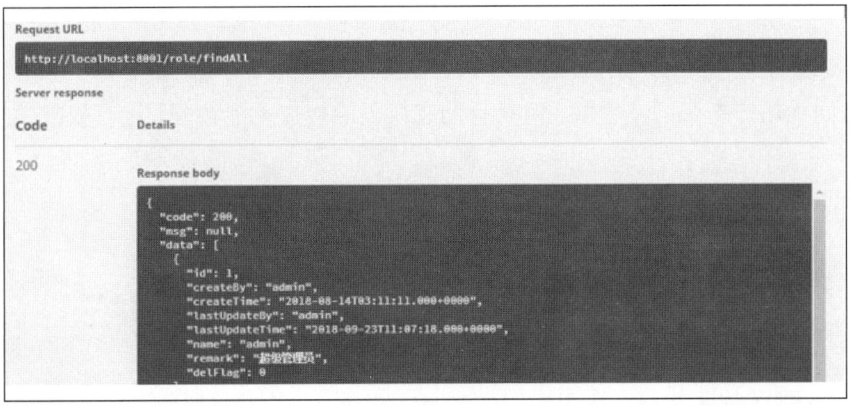

图 11-9

回到登录接口，换一个用户登录（"用户名：zhugeliang""密码：123"），因为这个用户有分配角色权限但没有分配菜单权限，所以如果权限注解接口保护正常，那么此用户访问角色接口正常（见图 11-10），但访问菜单相关接口会返回 403 错误（见图 11-11）。

图 11-10

图 11-11

虽然是同一用户携带同一令牌,但只有拥有权限的接口才能访问,没有权限的接口不允许访问。

# 11.5 Spring Security 执行流程剖析

Spring Security 功能强大,使用也稍显复杂,因为涉及的内容比较多,所以入门门槛也比较高,很多从业人员深受其烦,不少人就算跟着网上教程会用项目案例了,但是遇到问题还是摸不着头脑,这都是因为对其执行流程不熟悉造成的。

由于源码分析文章需要选取大量代码和图片,一是会占用大量篇幅,二是代码需要高亮标注的地方用纸质图书不好体现,因此这里附上本人博客的一篇剖析文章,通过追踪与解读源码的方式为读者详细剖析 Spring Security 的执行流程。熟悉整个流程之后,许多问题都可以迎刃而解了,有兴趣的读者可以自行查阅:https://www.cnblogs.com/xifengxiaoma/p/10020960.html。

# 第 12 章 数据备份还原

在很多时候，我们需要对系统数据进行备份还原。当然，实际生产环境的数据备份和还原通常是由专业数据库维护人员在数据库端通过命令执行的。这里提供的是通过代码进行数据备份，主要是方便一些日常的数据恢复，比如说想把数据恢复到某一世界节点的数据。这一章节，我们讲解如何通过代码调用 MySQL 的备份还原命令实现系统备份还原的功能。

## 12.1 新建工程

新建 mango-backup 工程，这是一个独立运行于 admin 的服务模块，可以分开独立部署，如图 12-1 所示。

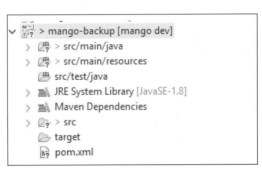

图 12-1

## 12.2 添加依赖

在 pom 文件中添加以下相关依赖，主要包含 web、swagger 和 common 依赖包。

pom.xml

```xml
<dependencies>
    <!-- spring boot -->
    <dependency>
        <groupId>org.springframework.boot</groupId>
```

```xml
        <artifactId>spring-boot-starter-web</artifactId>
    </dependency>
    <!-- swagger -->
    <dependency>
        <groupId>io.springfox</groupId>
        <artifactId>springfox-swagger2</artifactId>
        <version>2.9.2</version>
    </dependency>
    <dependency>
        <groupId>io.springfox</groupId>
        <artifactId>springfox-swagger-ui</artifactId>
        <version>2.9.2</version>
    </dependency>
    <dependency>
        <groupId>com.louis</groupId>
        <artifactId>mango-common</artifactId>
        <version>1.0.0</version>
    </dependency>
</dependencies>
```

## 12.3 添加配置

在配置文件中添加以下配置，定义启动端口为 8002、应用名称为 mango-backup，定义系统数据备份还原的数据库连接信息。

**application.yml**

```yml
# tomcat
server:
  port: 8002
spring:
  application:
    name: mango-backup
# backup datasource
mango:
  backup:
    datasource:
      host: localhost
      userName: root
      password: 123456
      database: mango
```

## 12.4 自定 Banner

在 resources 目录下添加一个自定义 banner.txt 文件，通过字符生成网站生成并输入以下内容：

```
    //   )  )                           //
   //__/ /     ___      ___     //  ___        ___
  / __  (    //   ) ) //   ) ) //\ \    //   / / //   ) )
 //    ) ) //   / / //       //  \ \  //   / / //___/ /
//____/ / ((___( ( ((____   //    \ \ ((___( ( //

========= Mango Application === Version: 1.0 =========
```

## 12.5 启动类

修改启动类为 MangoBackupApplication，指定包扫描路径为 com.louis.mango。

**MangoBackupApplication.java**

```java
@SpringBootApplication(scanBasePackages={"com.louis.mango"})
public class MangoBackupApplication {

    public static void main(String[] args) {
        SpringApplication.run(MangoBackupApplication.class, args);
    }

}
```

## 12.6 跨域配置

在 config 包下添加跨域配置类，前面章节已经提及，这里不再赘述。

**CorsConfig.java**

```java
public class CorsConfig implements WebMvcConfigurer {

    @Override
    public void addCorsMappings(CorsRegistry registry) {
```

```
        registry.addMapping("/**")    // 允许跨域访问的路径
                .allowedOrigins("*") // 允许跨域访问的源
                .allowedMethods("POST", "GET", "PUT", "OPTIONS", "DELETE")    // 允许请求方法
                .maxAge(168000)    // 预检间隔时间
                .allowedHeaders("*")    // 允许头部设置
                .allowCredentials(true); // 是否发送 cookie
    }

}
```

## 12.7　Swagger 配置

在 config 包下添加 Swagger 配置类，前面章节已经提及，这里不再赘述。

**SwaggerConfig.java**

```
@Configuration
@EnableSwagger2
public class SwaggerConfig {

    @Bean
    public Docket createRestApi() {
        return new Docket(DocumentationType.SWAGGER_2).select()

.apis(RequestHandlerSelectors.any()).paths(PathSelectors.any()).build();
    }

}
```

## 12.8　数据源属性

添加一个数据源属性配置类，配置@ConfigurationProperties(prefix = "mango.backup.datasource")注解，这样就可以通过注入 BackupDataSourceProperties 读取数据源属性了。

**BackupDataSourceProperties.java**

```
@Component
@ConfigurationProperties(prefix = "mango.backup.datasource")
public class BackupDataSourceProperties {
```

```
    private String host;
    private String userName;
    private String password;
    private String database;

    // 此处省略 getter 和 setter
}
```

## 12.9 备份还原接口

在 service 包下添加一个 MysqlBackupService 接口，包含 backup 和 restore 两个接口，分别对应备份和还原的接口。

**MysqlBackupService.java**

```
/**
 * MySQL 命令行备份恢复服务
 * @author Louis
 * @date Jan 15, 2019
 */
public interface MysqlBackupService {

    /**
     * 备份数据库
     * @param host host 地址，可以是本机也可以是远程
     * @param userName 数据库的用户名
     * @param password 数据库的密码
     * @param savePath 备份的路径
     * @param fileName 备份的文件名
     * @param databaseName 需要备份的数据库的名称
     * @return
     * @throws IOException
     */
    boolean backup(String host, String userName, String password,
    String backupFolderPath, String fileName, String database) throws Exception;

    /**
     * 恢复数据库
     * @param restoreFilePath 数据库备份的脚本路径
     * @param host IP 地址
     * @param database 数据库名称
     * @param userName 用户名
```

```
 * @param password 密码
 * @return
 */
boolean restore(String restoreFilePath, String host, String userName, String password,
    String database) throws Exception;
}
```

## 12.10 备份还原实现

在 service.impl 包下添加 MysqlBackupServiceImpl，实现 backup 和 restore 两个接口。

MysqlBackupServiceImpl.java

```
@Service
public class MysqlBackupServiceImpl implements MysqlBackupService {

    @Override
    public boolean backup(String host, String userName, String password, String backupFolderPath, String fileName,
            String database) throws Exception {
        return MySqlBackupRestoreUtils.backup(host, userName, password, backupFolderPath, fileName, database);
    }

    @Override
    public boolean restore(String restoreFilePath, String host, String userName, String password, String database)
            throws Exception {
        return MySqlBackupRestoreUtils.restore(restoreFilePath, host, userName, password, database);
    }
}
```

## 12.11 备份还原逻辑

为了方便复用，我们将系统数据备份和还原的逻辑封装到了 MySqlBackupRestoreUtils 工具类，主要是通过代码调用 MySQL 的数据库备份和还原命令实现，详见代码和注释。

## 12.11.1 数据备份服务

backup 是数据备份服务，读取数据源信息及备份信息生成数据备份。

MySqlBackupRestoreUtils.java

```java
/**
 * 备份数据库
 * @param host host 地址，可以是本机也可以是远程
 * @param userName 数据库的用户名
 * @param password 数据库的密码
 * @param savePath 备份的路径
 * @param fileName 备份的文件名
 * @param databaseName 需要备份的数据库的名称
 * @throws IOException
 */
public static boolean backup(String host, String userName, String password, String backupFolderPath, String fileName, String database) throws Exception {
    File backupFolderFile = new File(backupFolderPath);
    if (!backupFolderFile.exists()) {
        // 如果目录不存在则创建
        backupFolderFile.mkdirs();
    }
    if (!backupFolderPath.endsWith(File.separator)
            && !backupFolderPath.endsWith("/")) {
        backupFolderPath = backupFolderPath + File.separator;
    }
    // 拼接命令行的命令
    String backupFilePath = backupFolderPath + fileName;
    StringBuilder stringBuilder = new StringBuilder();
    stringBuilder.append("mysqldump --opt ").append(" --add-drop-database ").append(" --add-drop-table ");
    stringBuilder.append(" -h").append(host).append(" -u").append(userName).append(" -p").append(password);
    stringBuilder.append(" --result-file=").append(backupFilePath).append(" --default-character-set=utf8 ").append(database);
    // 调用外部执行 exe 文件的 Java API
    Process process = Runtime.getRuntime().exec(getCommand(stringBuilder.toString()));
    if (process.waitFor() == 0) {
        // 0 表示线程正常终止
        System.out.println("数据已经备份到 " + backupFilePath + " 文件中");
        return true;
    }
    return false;
}
```

### 12.11.2 数据还原服务

restore 是数据还原服务，读取数据源信息和还原版本进行数据还原 。

MySqlBackupRestoreUtils.java

```java
/**
 * 还原数据库
 * @param restoreFilePath 数据库备份的脚本路径
 * @param host IP 地址
 * @param database 数据库名称
 * @param userName 用户名
 * @param password 密码
 */
public static boolean restore(String restoreFilePath, String host, String userName,
String password, String database) throws Exception {
    File restoreFile = new File(restoreFilePath);
    if (restoreFile.isDirectory()) {
        for (File file : restoreFile.listFiles()) {
            if (file.exists() && file.getPath().endsWith(".sql")) {
                restoreFilePath = file.getAbsolutePath();
                break;
            }
        }
    }
    StringBuilder stringBuilder = new StringBuilder();
    stringBuilder.append("mysql -h").append(host).append(" -u").append(userName).append(" -p").append(password);
    stringBuilder.append(" ").append(database).append("<").append(restoreFilePath);
    try {
        Process process = Runtime.getRuntime().exec(getCommand
(stringBuilder.toString()));
        if (process.waitFor() == 0) {
            System.out.println("数据已从 " + restoreFilePath + " 导入到数据库中");
        }
    } catch (IOException e) {
        e.printStackTrace();
        return false;
    }
    return true;
}
```

## 12.12 备份还原控制器

在 controller 包下新建一个控制器，备份还原控制器对外提供数据备份还原的服务，包含以下几个接口。

### 12.12.1 数据备份接口

backup 是数据备份接口，读取数据源信息及备份信息生成数据备份。

MySqlBackupController.java

```java
@GetMapping("/backup")
public HttpResult backup() {
    String backupFodlerName = BackupConstants.DEFAULT_BACKUP_NAME + "_"
        + (new SimpleDateFormat(BackupConstants.DATE_FORMAT)).format(new Date());
    return backup(backupFodlerName);
}

private HttpResult backup(String backupFodlerName) {
    String host = properties.getHost();
    String userName = properties.getUserName();
    String password = properties.getPassword();
    String database = properties.getDatabase();
    String backupFolderPath = BackupConstants.BACKUP_FOLDER + backupFodlerName + File.separator;
    String fileName = BackupConstants.BACKUP_FILE_NAME;
    try {
        boolean success = mysqlBackupService.backup(host, userName, password, backupFolderPath, fileName, database);
        if(!success) {
            HttpResult.error("数据备份失败");
        }
    } catch (Exception e) {
        return HttpResult.error(500, e.getMessage());
    }
    return HttpResult.ok();
}
```

### 12.12.2 数据还原接口

restore 是数据还原接口，读取数据源信息和还原版本进行数据还原 。

**MySqlBackupController.java**

```java
@GetMapping("/restore")
public HttpResult restore(@RequestParam String name) throws IOException {
    String host = properties.getHost();
    String userName = properties.getUserName();
    String password = properties.getPassword();
    String database = properties.getDatabase();
    String restoreFilePath = BackupConstants.RESTORE_FOLDER + name;
    try {
        mysqlBackupService.restore(restoreFilePath, host, userName, password, database);
    } catch (Exception e) {
        return HttpResult.error(500, e.getMessage());
    }
    return HttpResult.ok();
}
```

### 12.12.3 查找备份接口

findRecords 是备份查找接口，用于查找和向用户展示数据备份版本。

**MySqlBackupController.java**

```java
@GetMapping("/findRecords")
public HttpResult findBackupRecords() {
    if(!new File(BackupConstants.DEFAULT_RESTORE_FILE).exists()) {
        // 初始默认备份文件
        backup(BackupConstants.DEFAULT_BACKUP_NAME);
    }
    List<Map<String, String>> backupRecords = new ArrayList<>();
    File restoreFolderFile = new File(BackupConstants.RESTORE_FOLDER);
    if(restoreFolderFile.exists()) {
        for(File file:restoreFolderFile.listFiles()) {
            Map<String, String> backup = new HashMap<>();
            backup.put("name", file.getName());
            backup.put("title", file.getName());

            if(BackupConstants.DEFAULT_BACKUP_NAME.equalsIgnoreCase(file.getName())) {
                backup.put("title", "系统默认备份");
            }
            backupRecords.add(backup);
        }
    }
```

```
    // 排序，默认备份最前，然后按时间戳排序，新备份在前面
    backupRecords.sort((o1, o2) ->
BackupConstants.DEFAULT_BACKUP_NAME.equalsIgnoreCase(o1.get("name")) ? -1
            :
BackupConstants.DEFAULT_BACKUP_NAME.equalsIgnoreCase(o2.get("name")) ? 1 :
o2.get("name").compareTo(o1.get("name")));
    return HttpResult.ok(backupRecords);
}
```

### 12.12.4　删除备份接口

delete 是备份删除接口，通过备份还原管理界面删除数据备份版本。

MySqlBackupController.java

```
@GetMapping("/delete")
public HttpResult deleteBackupRecord(@RequestParam String name) {
    if(BackupConstants.DEFAULT_BACKUP_NAME.equals(name)) {
        return HttpResult.error("系统默认备份无法删除!");
    }
    String restoreFilePath = BackupConstants.BACKUP_FOLDER + name;
    try {
        FileUtils.deleteFile(new File(restoreFilePath));
    } catch (Exception e) {
        return HttpResult.error(500, e.getMessage());
    }
    return HttpResult.ok();
}
```

## 12.13　接口测试

编译启动应用，运行 MangoBackupApplication，访问 http://localhost:8002/swagger-ui.html#/，进入 Swagger 接口测试页面，如图 12-2 所示。

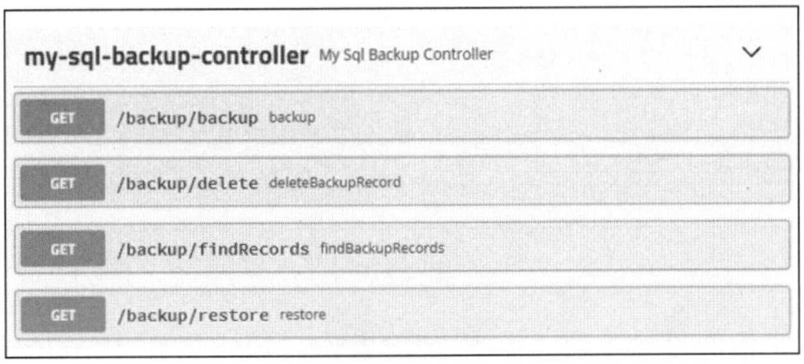

图 12-2

首先，调用以下 findRecords 接口，发现返回了一条"系统默认备份"数据，这是因为我们为了保证系统至少保留一条备份而自动生成的备份，生成逻辑是每次在查找备份的时候，如果发现系统不存在"系统默认备份"，就会利用当前数据库数据生成一条"系统默认备份"，如图 12-3 所示。

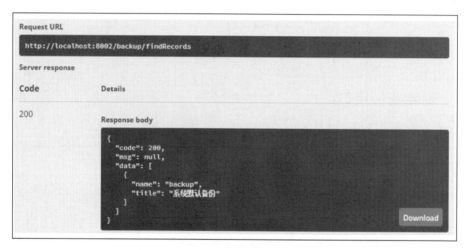

图 12-3

生成备份需要存储成 SQL 文件，备份文件的存储目录是：系统用户目录 / _mango_backup。在 _mango_backup 目录下会存储很多目录，每个目录代码一个备份版本，目录下面就是该版本的 SQL 文件。"系统默认备份"目录名为"mango"，普通备份目录名为"mango_时间戳"。

自动生成的"系统默认备份"版本文件如图 12-4 所示。

图 12-4

接着调用一下备份接口，果不其然，生成了一份"mango_时间戳"的备份目录，如图 12-5 所示。

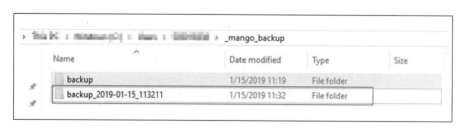

图 12-5

然后输入新生成的备份目录名称，调用删除备份接口，如图 12-6 所示。

查看备份存储目录，发现刚才新生成的备份目录已经成功被删除了，如图 12-7 所示。

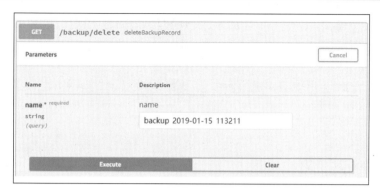

图 12-6

图 12-7

现在，我们来测试一下数据还原，先把数据里的表全部删除，通过命令行或工具都行，如图 12-8 所示。

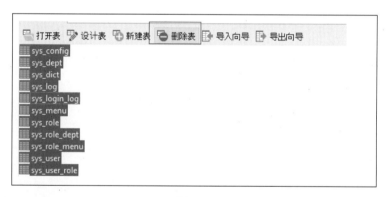

图 12-8

然后找到数据还原接口，输入备份目录名称，比如系统默认备份是 backup，如图 12-9 所示。

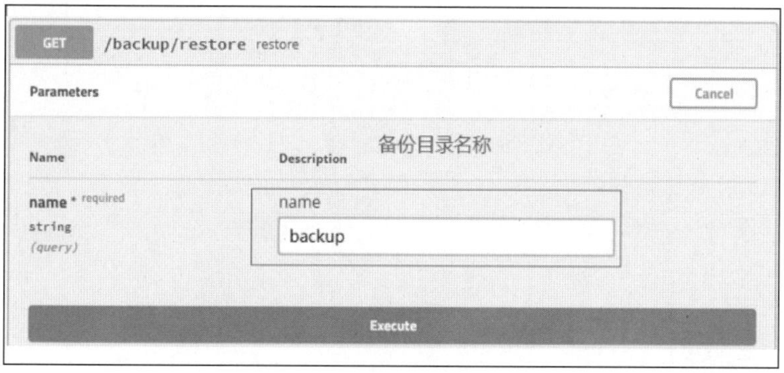

图 12-9

执行成功之后，查看并刷新一下数据库，发现被我们删除的数据库表和数据都成功恢复回来了，如图 12-10 所示。

图 12-10

# 第 13 章 系统服务监控

Spring Boot Admin 是一个管理和监控 Spring Boot 应用程序的开源监控软件，针对 spring-boot 的 actuator 接口进行 UI 美化并封装,可以在管理界面中浏览所有被监控 spring-boot 项目的基本信息，详细的 Health 信息、内存信息、JVM 信息、垃圾回收信息、各种配置信息（比如数据源、缓存列表和命中率）等，还可以直接修改 logger 的 level，Spring Boot Admin 提供的丰富详细的监控信息给 Spring Boot 应用的监控、维护和优化都带来了极大的便利。

本章就给大家介绍如何使用 Spring Boot Admin 对 Spring Boot 应用进行监控。

## 13.1 新建工程

新建一个 mango-monitor 项目，作为服务监控服务端，工程结构如图 13-1 所示。

图 13-1

## 13.2 添加依赖

在 pom 文件中添加以下依赖包，主要是 Spring Boot 和 Spring Boot Admin 依赖。

### pom.xml

```
<!-- spring boot -->
<dependency>
```

```xml
    <groupId>org.springframework.boot</groupId>
    <artifactId>spring-boot-starter</artifactId>
</dependency>
<!--spring-boot-admin-->
<dependency>
    <groupId>de.codecentric</groupId>
    <artifactId>spring-boot-admin-server</artifactId>
    <version>2.1.2</version>
</dependency>
<dependency>
    <groupId>de.codecentric</groupId>
    <artifactId>spring-boot-admin-server-ui</artifactId>
    <version>2.1.2</version>
</dependency>
```

## 13.3 添加配置

在配置文件中添加配置，指定启动端口为 8000、应用名称为 mango-monitor。

application.yml

```yml
server:
  port: 8000
spring:
  application:
    name: mango-monitor
```

## 13.4 自定义 Banner

在 resources 目录下添加 banner.txt 文件，生成 banner 字符并写入。

```
   /|    //| |
  //|   // | |     __       __      (  )  __   __   __    __
 // |  //  | |    //  ) )  //  ) ) / /   / /  //  ) ) //  ) )
//  | //   | |   //      //   / / / /   / /  //      //   / /
//   |//    | | ((__ / / //   / / / /   ((__/ /  //      ((__/ /
========= Mango Application === Version: 1.0 =========
```

## 13.5 启动类

修改启动类为 MangoMonitorApplication 并在头部添加@EnableAdminServer 注解,开启监控服务。

MangoMonitorApplication.java

```java
@EnableAdminServer
@SpringBootApplication
public class MangoMonitorApplication {

    public static void main(String[] args) {
        SpringApplication.run(MangoMonitorApplication.class, args);
    }

}
```

## 13.6 启动服务端

编译启动 MangoMonitorApplication,访问 http://localhost:8000,进入应用监控界面,如图 13-2 所示。

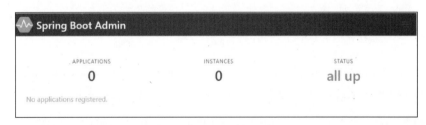

图 13-2

## 13.7 监控客户端

现在管理界面应用和实例都是 0,因为还没有监控客户端被注册,接下来我们就把后台服务 mango-admin 和数据备份还原服务 mango-backup 注册到监控服务。

分别在 mango-admin 和 mango-backup 的 pom 文件中添加监控客户端依赖。

pom.xml

```xml
<!--spring-boot-admin-client-->
<dependency>
    <groupId>de.codecentric</groupId>
    <artifactId>spring-boot-admin-starter-client</artifactId>
    <version>2.1.2</version>
</dependency>
```

分别在 mango-admin 和 mango-backup 的配置文件中配置服务监控信息。主要是指定监控服务器的地址，另外 endpoints 是开放监控接口，因为处于安全的考虑，Spring Boot 默认是没有开放健康检查接口的，可以通过 endpoints 设置开放特定的接口，"*"表示全部开放。

application.yml

```yml
spring:
  boot:
    admin:
      client:
        url: "http://localhost:8000"
management:
  endpoints:
    web:
      exposure:
        include: "*"
```

## 13.8 启动客户端

客户端添加服务监控之后，分别编译启动 mang-admin 和 mango-backup。注意，在此之前请保证 mango-monitor 服务已经启动。客户端启动之后，服务监控界面如图 13-3 所示。

图 13-3

Wallboard 页面按六边形块展示当前监控的应用，如图 13-4 所示。

图 13-4

单击应用进入应用监控详情页面，我们可以看到各种应用监控信息，比如应用环境、处理器信息、线程信息、类加载、堆栈使用和垃圾回收情况等，如图 13-5 所示。

图 13-5

另外，还有诸如服务上线日志等，功能丰富，有兴趣的读者可自行探究，如图 13-6 所示。

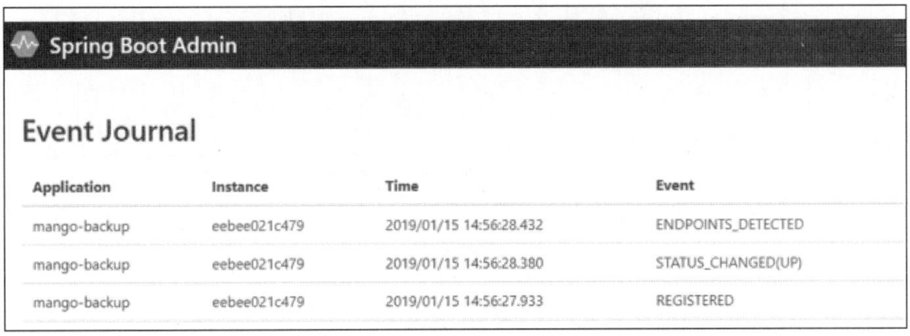

图 13-6

# 第 14 章 注册中心（Consul）

## 14.1 什么是 Consul

Consul 是 HashiCorp 公司推出的开源工具，用于实现分布式系统的服务发现与配置。与其他分布式服务注册与发现的方案相比，Consul 的方案更"一站式"，内置了服务注册与发现框架、分布一致性协议实现、健康检查、Key/Value 存储、多数据中心方案，不再需要依赖其他工具（比如 ZooKeeper 等）。使用起来也较为简单。Consul 使用 Go 语言编写，因此具有天然可移植性（支持 Linux、Windows 和 Mac OS X）；安装包仅包含一个可执行文件，方便部署，与 Docker 等轻量级容器可无缝配合。

## 14.2 Consul 安装

访问 Consul 官网，根据操作系统类型，选择下载 Consul 的最新版本。这里选择 Windows 1.4.0 版本，如图 14-1 所示。下载地址为 https://www.consul.io/downloads.html。

下载下来是一个 zip 压缩包，解压之后，是一个 exe 可执行文件，如图 14-2 所示。

图 14-1

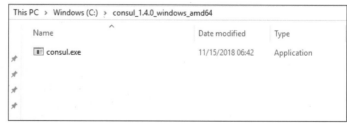

图 14-2

打开 CMD 终端，进入 consul.exe 所在目录，执行如下命令启动 Consul 服务。

```
# 进入consul.exe所在目录
cd C:\consul_1.4.0_windows_amd64
```

```
# 启动服务，-dev 表示开发模式运行，另外还有 -server 表示服务模式运行
consul agent -dev
```

启动过程信息如图 14-3 所示。

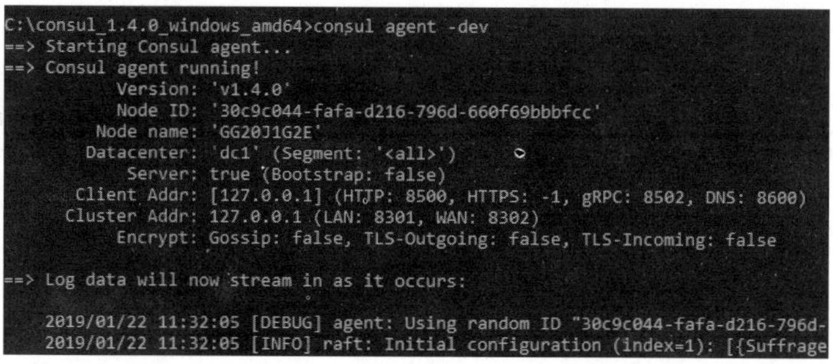

图 14-3

启动成功之后，访问 http://localhost:8500，如果可以看到如图 14-4 所示的 Consul 服务管理界面，就说明注册中心服务端可以正常提供服务了。

图 14-4

# 14.3 monitor 改造

改造 mango-monitor 工程，作为服务注册到注册中心。

## 14.3.1 添加依赖

在 pom.xml 中添加 Spring Cloud 和 Consul 注册中心依赖。

注意

Spring Boot 2.1 后的版本会出现 Consul 服务注册上的问题，可能是因为配置变更或者支持方式发生改变，由于版本太新，网上也没有相关解决方案，所以这里把 Spring Boot 版本调整为 2.0.4，Spring Cloud 版本使用最新的稳定发布版 Finchley.RELEASE。

### Consul 依赖

```xml
<!--consul-->
<dependency>
    <groupId>org.springframework.cloud</groupId>
    <artifactId>spring-cloud-starter-consul-discovery</artifactId>
</dependency>
```

### Spring Cloud 依赖

```xml
<!--srping cloud-->
<dependencyManagement>
    <dependencies>
        <dependency>
            <groupId>org.springframework.cloud</groupId>
            <artifactId>spring-cloud-dependencies</artifactId>
            <version>Finchley.RELEASE</version>
            <type>pom</type>
            <scope>import</scope>
        </dependency>
    </dependencies>
</dependencyManagement>
```

## 14.3.2 配置文件

修改配置文件，添加服务注册配置。

### application.yml

```yaml
server:
  port: 8000
spring:
  application:
    name: mango-monitor
  cloud:
    consul:
      host: localhost
      port: 8500
      discovery:
        serviceName: ${spring.application.name}   # 注册到 consul 的服务名称
```

## 14.3.3 启动类

修改启动类，添加 @EnableDiscoveryClient 注解，开启服务发现支持。

MangoMonitorApplication.java

```
@EnableAdminServer
@EnableDiscoveryClient
@SpringBootApplication
public class MangoMonitorApplication {

    public static void main(String[] args) {
        SpringApplication.run(MangoMonitorApplication.class, args);
    }

}
```

## 14.3.4 测试效果

启动服务监控服务器，访问 http://localhost:8500，发现服务已经成功注册到注册中心，如图 14-5 所示。

图 14-5

访问 http://localhost:8000，查看服务监控管理界面，看到如图 14-6 所示的界面就没问题了。

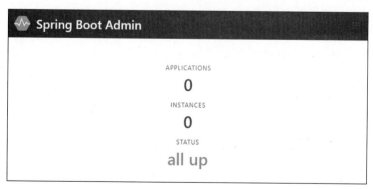

图 14-6

## 14.4 backup 改造

改造 mango-backup 工程，作为服务注册到注册中心。

### 14.4.1 添加依赖

在 pom.xml 中添加 Spring Cloud 和 Consul 注册中心依赖。

**Consul 依赖**

```xml
<!--consul-->
<dependency>
    <groupId>org.springframework.cloud</groupId>
    <artifactId>spring-cloud-starter-consul-discovery</artifactId>
</dependency>
```

**Spring Cloud 依赖**

```xml
<!--srping cloud-->
<dependencyManagement>
    <dependencies>
        <dependency>
            <groupId>org.springframework.cloud</groupId>
            <artifactId>spring-cloud-dependencies</artifactId>
            <version>Finchley.RELEASE</version>
            <type>pom</type>
            <scope>import</scope>
        </dependency>
    </dependencies>
</dependencyManagement>
```

### 14.4.2 配置文件

修改配置文件，添加服务注册配置。

**application.yml**

```yml
# tomcat
server:
  port: 8002
spring:
  application:
    name: mango-backup
```

```yaml
  boot:
    admin:
      client:
        url: "http://localhost:8000"
  cloud:
    consul:
      host: localhost
      port: 8500
      discovery:
        serviceName: ${spring.application.name} # 注册到 consul 的服务名称
# 开放健康检查接口
management:
  endpoints:
    web:
      exposure:
        include: "*"
  endpoint:
    health:
      show-details: ALWAYS
# backup datasource
mango:
  backup:
    datasource:
      host: localhost
      userName: root
      password: 123456
      database: mango
```

### 14.4.3 启动类

修改启动类，添加 @EnableDiscoveryClient 注解，开启服务发现支持。

MangoBackupApplication.java

```java
@EnableDiscoveryClient
@SpringBootApplication(scanBasePackages={"com.louis.mango"})
public class MangoBackupApplication {

    public static void main(String[] args) {
        SpringApplication.run(MangoBackupApplication.class, args);
    }

}
```

### 14.4.4 测试效果

启动备份服务，访问 http://localhost:8500，发现服务已经成功注册到注册中心，如图 14-7 所示。

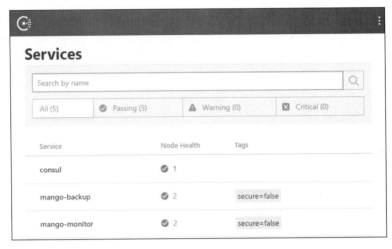

图 14-7

访问 http://localhost:8000，查看服务监控管理界面，发现服务已经在监控列表里了，如图 14-8 所示。

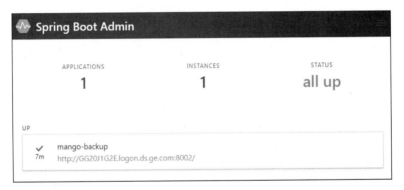

图 14-8

## 14.5 admin 改造

改造 mango-admin 工程，作为服务注册到注册中心。

### 14.5.1 添加依赖

在 pom.xml 中添加 Spring Cloud 和 Consul 注册中心依赖。

### Consul 依赖

```xml
<!--consul-->
<dependency>
    <groupId>org.springframework.cloud</groupId>
    <artifactId>spring-cloud-starter-consul-discovery</artifactId>
</dependency>
```

### Spring Cloud 依赖

```xml
<!--srping cloud-->
<dependencyManagement>
    <dependencies>
        <dependency>
            <groupId>org.springframework.cloud</groupId>
            <artifactId>spring-cloud-dependencies</artifactId>
            <version>Finchley.RELEASE</version>
            <type>pom</type>
            <scope>import</scope>
        </dependency>
    </dependencies>
</dependencyManagement>
```

## 14.5.2 配置文件

修改配置文件，添加服务注册配置。

### application.yml

```yml
server:
  port: 8001
spring:
  application:
    name: mango-admin
  boot:
    admin:
      client:
        url: "http://localhost:8000"
  datasource:
    name: druidDataSource
    type: com.alibaba.druid.pool.DruidDataSource
    druid:
      driver-class-name: com.mysql.jdbc.Driver
      url: jdbc:mysql://localhost:3306/mango?useUnicode=true&zeroDateTimeBehavior=convertToNull&autoReconnect=true&characterEncoding=utf-8
      username: root
      password: 123456
```

```yaml
        filters: stat,wall,log4j,config
        max-active: 100
        initial-size: 1
        max-wait: 60000
        min-idle: 1
        time-between-eviction-runs-millis: 60000
        min-evictable-idle-time-millis: 300000
        validation-query: select 'x'
        test-while-idle: true
        test-on-borrow: false
        test-on-return: false
        pool-prepared-statements: true
        max-open-prepared-statements: 50
        max-pool-prepared-statement-per-connection-size: 20
  cloud:
    consul:
      host: localhost
      port: 8500
      discovery:
        serviceName: ${spring.application.name}  # 注册到 consul 的服务名称
# 开放健康检查接口
management:
  endpoints:
    web:
      exposure:
        include: "*"
  endpoint:
    health:
      show-details: ALWAYS
```

### 14.5.3 启动类

修改启动类，添加 @EnableDiscoveryClient 注解，开启服务发现支持。

MangoAdminApplication.java

```java
@EnableDiscoveryClient
@SpringBootApplication(scanBasePackages={"com.louis.mango"})
public class MangoAdminApplication {

    public static void main(String[] args) {
        SpringApplication.run(MangoAdminApplication.class, args);
    }

}
```

### 14.5.4 测试效果

启动服务监控服务器，访问 http://localhost:8500，发现服务已经成功注册到注册中心，如图 14-9 所示。

图 14-9

访问 http://localhost:8000，查看服务监控管理界面，发现服务已经在监控列表里了，如图 14-10 所示。

图 14-10

# 第 15 章 服务消费（Ribbon、Feign）

## 15.1 技术背景

在第 14 章中，我们利用 Consul 注册中心实现了服务的注册和发现功能，这一章我们来聊聊服务的调用。在单体应用中，代码可以直接依赖，在代码中直接调用即可；但在微服务架构（分布式架构）中，服务都运行在各自的进程之中，甚至部署在不同的主机和不同的地区，就需要相关的远程调用技术了。

Spring Cloud 体系里应用比较广泛的服务调用方式有两种：

（1）使用 RestTemplate 进行服务调用，可以通过 Ribbon 注解 RestTemplate 模板，使其拥有负载均衡的功能。

（2）使用 Feign 进行声明式服务调用，声明之后就像调用本地方法一样，Feign 默认使用 Ribbon 实现负载均衡。

两种方式都可以实现服务之间的调用，可根据情况选择使用，下面我们分别用实现案例来进行讲解。

## 15.2 服务提供者

### 15.2.1 新建项目

新建一个项目 mango-producer，添加以下依赖。

- Swagger：API 文档。
- Consul：注册中心。
- Spring Boot Admin：服务监控。

pom.xml

```xml
<dependency>
    <groupId>org.springframework.boot</groupId>
```

```xml
        <artifactId>spring-boot-starter-web</artifactId>
</dependency>
<!-- swagger -->
<dependency>
    <groupId>io.springfox</groupId>
    <artifactId>springfox-swagger2</artifactId>
    <version>${swagger.version}</version>
</dependency>
<dependency>
    <groupId>io.springfox</groupId>
    <artifactId>springfox-swagger-ui</artifactId>
    <version>${swagger.version}</version>
</dependency>
<!--spring-boot-admin-->
<dependency>
    <groupId>de.codecentric</groupId>
    <artifactId>spring-boot-admin-starter-client</artifactId>
    <version>${spring.boot.admin.version}</version>
</dependency>
<!--consul-->
<dependency>
    <groupId>org.springframework.cloud</groupId>
    <artifactId>spring-cloud-starter-consul-discovery</artifactId>
</dependency>
```

### 15.2.2 配置文件

在配置文件添加内容如下，将服务注册到注册中心并添加服务监控相关配置。

application.yml

```yaml
server:
  port: 8003
spring:
  application:
    name: mango-producer
  cloud:
    consul:
      host: localhost
      port: 8500
      discovery:
        serviceName: ${spring.application.name} # 注册到 consul 的服务名称
  boot:
    admin:
```

```yaml
    client:
      url: "http://localhost:8000"
# 开放健康检查接口
management:
  endpoints:
    web:
      exposure:
        include: "*"
  endpoint:
    health:
      show-details: ALWAYS
```

### 15.2.3 启动类

修改启动器类，添加 @EnableDiscoveryClient 注解，开启服务发现支持。

MangoProducerApplication.java

```java
@EnableDiscoveryClient
@SpringBootApplication
public class MangoProducerApplication {

    public static void main(String[] args) {
        SpringApplication.run(MangoProducerApplication.class, args);
    }

}
```

### 15.2.4 自定义 Banner

在 resources 目录下新建一个自定义 banner 文件。

```
banner.txt
  _____        ____    __| _/_  __    ____    _____
 /  \__  \ /    \  / _ \ /    |   | \/ ___\ /  _ \   \____ \
|   | ___/|   |  \(  <_> )/_/ |   | /\  \__(  <_> )  |  |_> >
|___|  /  |___|  / \____/ \____ |___|/  \___  >\____/   |   __/
     \/        \/              \/           \/          |__|
========= Mango Application === Version: 1.0 =========
```

### 15.2.5 添加控制器

新建一个控制器，提供一个 hello 接口，返回字符串信息。

HelloController.java

```java
@RestController
public class HelloController {

    @RequestMapping("/hello")
    public String hello() {
        return "hello Mango !";
    }
}
```

为了模拟均衡负载，复制一份上面的项目，重命名为 mango-producer2，修改对应的端口为 8004，修改 hello 方法的返回值为："hello Manog 2!"。

依次启动注册中心、服务监控和两个服务提供者，启动成功之后刷新 Consul 管理界面，发现我们注册的 mango-producer 服务以及有 2 个节点实例。

访问 http://localhost:8500，查看两个服务提供者已经注册到注册中心，如图 15-1 所示。

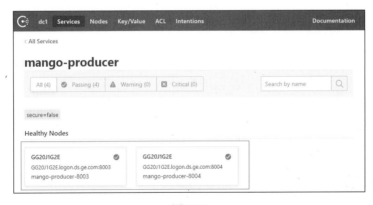

图 15-1

访问 http://localhost:8000，查看两个服务提供者已经成功显示在监控列表中，如图 15-2 所示。

图 15-2

访问 http://localhost:8003/hello，输出"hello Mango!"，如图 15-3 所示。

图 15-3

访问 http://localhost:8004/hello，输出"hello Mango2!"，如图 15-4 所示。

图 15-4

# 15.3 服务消费者

## 15.3.1 新建项目

新建一个项目 mango-consumer，添加以下依赖。

- Swagger：API 文档。
- Consul：注册中心。
- Spring Boot Admin：服务监控。

pom.xml

```
<!-- web -->
<dependency>
    <groupId>org.springframework.boot</groupId>
    <artifactId>spring-boot-starter-web</artifactId>
</dependency>
<!-- swagger -->
<dependency>
    <groupId>io.springfox</groupId>
    <artifactId>springfox-swagger2</artifactId>
    <version>${swagger.version}</version>
</dependency>
<dependency>
    <groupId>io.springfox</groupId>
    <artifactId>springfox-swagger-ui</artifactId>
```

```xml
    <version>${swagger.version}</version>
</dependency>
<!--spring-boot-admin-->
  <dependency>
    <groupId>de.codecentric</groupId>
    <artifactId>spring-boot-admin-starter-client</artifactId>
    <version>${spring.boot.admin.version}</version>
</dependency>
<!--consul-->
<dependency>
    <groupId>org.springframework.cloud</groupId>
    <artifactId>spring-cloud-starter-consul-discovery</artifactId>
</dependency>
```

### 15.3.2 添加配置

修改配置文件如下。

application.yml

```yml
server:
 port: 8005
spring:
 application:
   name: mango-consumer
 cloud:
   consul:
     host: localhost
     port: 8500
     discovery:
       serviceName: ${spring.application.name}      # 注册到 consul 的服务名称
 boot:
   admin:
     client:
       url: "http://localhost:8000"
# 开放健康检查接口
management:
  endpoints:
    web:
      exposure:
        include: "*"
  endpoint:
    health:
      show-details: ALWAYS
```

### 15.3.3 启动类

修改启动器类，添加 @EnableDiscoveryClient 注解，开启服务发现支持。

MangoConsumerApplication.java

```java
@EnableDiscoveryClient
@SpringBootApplication
public class MangoConsumerApplication {

    public static void main(String[] args) {
        SpringApplication.run(MangoConsumerApplication.class, args);
    }

}
```

### 15.3.4 自定义 Banner

在 resources 目录下新建一个自定义 banner 文件。

banner.txt

```
    ___        ___         ___     ___     ___      _____
  _/ __\/   _ \ /    \ /  ___/ |  \/   \_/  __ \_   ___  \
  \  \_(  <_> )   |  |\___  \|  |  /  Y Y  \  ___/|    |\/
   \__  >____/|__|  /____  >___/|__|_|  /\___  >__|
       \/       \/       \/           \/     \/

========= Mango Application === Version: 1.0 ==========
```

### 15.3.5 服务消费

添加消费服务测试类，添加两个接口，一个查询所有我们注册的服务，另一个从我们注册的服务中选取一个服务，采用轮询的方式。

ServiceController.java

```java
@RestController
public class ServiceController {

    @Autowired
    private LoadBalancerClient loadBalancerClient;
    @Autowired
    private DiscoveryClient discoveryClient;
```

```
/**
 * 获取所有服务
 */
@RequestMapping("/services")
public Object services() {
    return discoveryClient.getInstances("mango-producer");
}

/**
 * 从所有服务中选择一个服务（轮询）
 */
@RequestMapping("/discover")
public Object discover() {
    return loadBalancerClient.choose("mango-producer").getUri().toString();
}
}
```

添加完成之后，启动项目，访问 http://localhost:8500，服务消费者已经成功注册到注册中心，如图 15-5 所示。

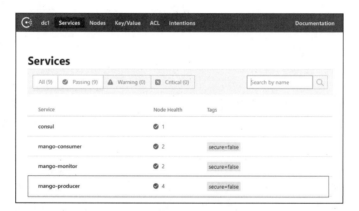

图 15-5

访问 http://localhost:8000，服务消费者已经成功显示在监控列表中，如图 15-6 所示。

图 15-6

访问 http://localhost:8005/services，返回两个服务，分别是我们注册的 8003 和 8004，如图 15-7 所示。

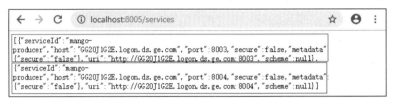

图 15-7

反复访问 http://localhost:8005/discover，结果交替返回服务 8003 和 8004，因为默认的负载均衡器采用的是轮询方式，如图 15-8 所示。8003 和 8004 两个服务交替出现，从而实现了获取服务端地址的均衡负载。

图 15-8

大多数情况下我们希望使用均衡负载的形式去获取服务端提供的服务，因此使用第二种方法来模拟调用服务端提供的 hello 方法。创建一个控制器 CallHelloController。

CallHelloController.java

```
@RestController
public class CallHelloController {

    @Autowired
    private LoadBalancerClient loadBalancer;

    @RequestMapping("/call")
    public String call() {
        ServiceInstance serviceInstance = loadBalancer.choose("mango-producer");
        System.out.println("服务地址: " + serviceInstance.getUri());
        System.out.println("服务名称: " + serviceInstance.getServiceId());

        String callServiceResult = new RestTemplate().getForObject(serviceInstance.getUri().toString() + "/hello", String.class);
        System.out.println(callServiceResult);
        return callServiceResult;
    }
}
```

使用 RestTemplate 进行远程调用。添加完之后重启 kitty-consumer 项目。在浏览器中访问 http://localhost:8005/call，依次往复返回的结果如图 15-9 所示。

图 15-9

## 15.3.6 负载均衡器（Ribbon）

在上面的教程中，我们是这样调用服务的，先通过 LoadBalancerClient 选取出对应的服务，然后使用 RestTemplate 进行远程调用。

LoadBalancerClient 就是负载均衡器，RibbonLoadBalancerClient 是 Ribbon 默认使用的负载均衡器，采用的负载均衡策略是轮询。

（1）查找服务，通过 LoadBalancer 查询服务。

```
ServiceInstance serviceInstance = loadBalancer.choose("mango-producer");
```

（2）调用服务，通过 RestTemplate 远程调用服务。

```
String callServiceResult = new
RestTemplate().getForObject(serviceInstance.getUri().toString() + "/hello",
String.class);
```

这样就完成了一个简单的服务调用和负载均衡。接下来我们说说 Ribbon。

Ribbon 是 Netflix 发布的负载均衡器，它有助于控制 HTTP 和 TCP 的客户端的行为。为 Ribbon 配置服务提供者地址后，Ribbon 就可基于某种负载均衡算法自动地帮助服务消费者去请求。Ribbon 默认为我们提供了很多负载均衡算法，例如轮询、随机等。当然，我们也可为 Ribbon 实现自定义的负载均衡算法。

Ribbon 内置负载均衡策略如表 15-1 所示。

表 15-1 ribbon 内置负载均衡策略

| 策略名 | 策略声明 | 策略描述 | 实现说明 |
| --- | --- | --- | --- |
| BestAvailableRule | public class BestAvailableRule extends ClientConfigEnabled RoundRobinRule | 选择一个最小的并发请求的 server | 逐个考察 Server，如果 Server 被 tripped 了就忽略，再选择其中 ActiveRequestsCount 最小的 server |
| AvailabilityFilteringRule | public class AvailabilityFilteringRule extends PredicateBasedRule | 过滤掉那些因为一直连接失败的被标记为 circuit tripped 的后端 server，并过滤掉那些高并发的的后端 server（active connections 超过配置的阈值） | 使用一个 AvailabilityPredicate 来包含过滤 server 的逻辑，其实就是检查 status 里记录的各 server 的运行状态 |

(续表)

| 策略名 | 策略声明 | 策略描述 | 实现说明 |
| --- | --- | --- | --- |
| WeightedResponseTimeRule | public class WeightedResponseTimeRule extends RoundRobinRule | 根据响应时间分配一个 weight,响应时间越长, weight 越小,被选中的可能性越低 | 一个后台线程定期地从 status 里面读取评价响应时间,为每个 server 计算一个 weight。weight 的计算比较简单,responsetime 减去每个 server 自己平均的 responsetime 是 server 的权重。当刚开始运行还没有形成 status 时,使用 roubine 策略选择 server |
| RetryRule | public class RetryRule extends AbstractLoadBalancerRule | 对选定的负载均衡策略机采取重试机制 | 在一个配置时间段内当选择 server 不成功时,就一直尝试使用 subRule 的方式选择一个可用的 server |
| RoundRobinRule | public class RoundRobinRule extends AbstractLoadBalancerRule | 使用 roundRobin 方式轮询选择 server | 轮询 index,选择 index 对应位置的 server |
| RandomRule | public class RandomRule extends AbstractLoadBalancerRule | 随机选择一个 server | 随机选择 index,再选择对应位置的 server |
| ZoneAvoidanceRule | public class ZoneAvoidanceRule extends PredicateBasedRule | 复合判断 server 所在区域的性能和 server 的可用性,根据结果来选择 server | 使用 ZoneAvoidancePredicate 和 AvailabilityPredicate 来判断是否选择某个 server,前一个判断判定一个 zone 的运行性能是否可用,剔除不可用的 zone 的所有 server,AvailabilityPredicate 用于过滤掉连接数过多的 server |

### 15.3.7 修改启动类

我们修改一下启动器类,注入 RestTemplate,并添加@LoadBalanced 注解(用于拦截请求),以使用 ribbon 来进行负载均衡。

MangoConsumerApplication.java

```
@EnableDiscoveryClient
@SpringBootApplication
public class MangoConsumerApplication {

    public static void main(String[] args) {
        SpringApplication.run(MangoConsumerApplication.class, args);
    }

    @Bean
```

```
    @LoadBalanced
    public RestTemplate restTemplate() {
        return new RestTemplate();
    }
}
```

### 15.3.8 添加服务

新建一个控制器类，注入 RestTemplate，并调用服务提供者的 hello 服务。

RibbonHelloController.java

```
@RestController
public class RibbonHelloController {

    @Autowired
    private RestTemplate restTemplate;

    @RequestMapping("/ribbon/call")
    public String call() {
        // 调用服务，service-producer 为注册的服务名称
        // LoadBalancerInterceptor 会拦截调用并根据服务名找到对应的服务
        String callServiceResult = restTemplate.getForObject("http://mango-producer/hello", String.class);
        return callServiceResult;
    }
}
```

### 15.3.9 页面测试

启动消费者服务，访问 http://localhost:8005/ribbon/call，依次返回的结果如图 15-10 所示。

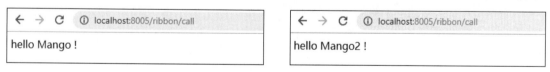

图 15-10

说明 ribbon 的负载均衡已经成功启动了。

### 15.3.10 负载策略

修改负载均衡策略很简单，只需要在配置文件指定对应的负载均衡器即可。如这里把策略修改为随机策略，之后负载均衡器会随机选取注册的服务。

application.yml

```yaml
#ribbon 负载均衡策略配置，service-producer 为注册的服务名
service-producer:
  ribbon:
    NFLoadBalancerRuleClassName: com.netflix.loadbalancer.RandomRule
```

## 15.4 服务消费（Feign）

Spring Cloud Feign 是一套基于 Netflix Feign 实现的声明式服务调用客户端，使编写 Web 服务客户端变得更加简单。我们只需要通过创建接口并用注解来配置它即可完成对 Web 服务接口的绑定。它具备可插拔的注解支持，包括 Feign 注解、JAX-RS 注解。它也支持可插拔的编码器和解码器。Spring Cloud Feign 还扩展了对 Spring MVC 注解的支持，同时还整合了 Ribbon 来提供均衡负载的 HTTP 客户端实现。

### 15.4.1 添加依赖

修改 mango-consumer 的 pom 文件，添加 feign 依赖。

pom.xml

```xml
<!--feign -->
<dependency>
    <groupId>org.springframework.cloud</groupId>
    <artifactId>spring-cloud-starter-openfeign</artifactId>
</dependency>
```

### 15.4.2 启动类

修改启动器类，添加 @EnableFeignClients 注解，开启扫描 Spring Cloud Feign 客户端的功能。修改完成之后的启动类如下。

MangoConsumerApplication.java

```java
@EnableFeignClients
@EnableDiscoveryClient
@SpringBootApplication
public class MangoConsumerApplication {

    public static void main(String[] args) {
        SpringApplication.run(MangoConsumerApplication.class, args);
    }

    @Bean
    @LoadBalanced
```

```
    public RestTemplate restTemplate() {
        return new RestTemplate();
    }
}
```

### 15.4.3 添加 Feign 接口

添加 MangoProducerService 接口，在类头添加注解 @FeignClient("mango-producer")，mango-producer 是要调用的服务名。

添加跟调用目标方法一样的方法声明，只需要方法声明，不需要具体实现，注意跟目标方法定义保持一致。

MangoProducerService.java

```
@FeignClient(name = "mango-producer")
public interface MangoProducerService {

    @RequestMapping("/hello")
    public String hello();

}
```

### 15.4.4 添加控制器

添加 FeignHelloController 控制器，注入 MangoProducerService，就可以像使用本地方法一样进行调用了。

FeignHelloController.java

```
@RestController
public class FeignHelloController {

    @Autowired
    private MangoProducerService mangoProducerService;

    @RequestMapping("/feign/call")
    public String call() {
        // 像调用本地服务一样
        return mangoProducerService.hello();
    }

}
```

### 15.4.5 页面测试

启动成功之后，访问 http://localhost:8005/feign/call，发现调用成功，且依次往复返回如图 15-11 所示的结果。

图 15-11

Feign 是声明式调用，会产生一些相关的 Feign 定义接口，所以建议将 Feign 定义的接口都统一放置管理，以区别内部服务。

# 第 16 章
# 服务熔断（Hystrix、Turbine）

## 16.1 雪崩效应

在微服务架构中，服务众多，通常会涉及多个服务层级的调用，一旦基础服务发生故障，很可能会导致级联故障，进而造成整个系统不可用，这种现象被称为服务雪崩效应。服务雪崩效应是一种因"服务提供者"的不可用导致"服务消费者"的不可用并将这种不可用逐渐放大的过程。

比如在一个系统中，A 是服务提供者，B 是 A 的服务消费者，C 和 D 又是 B 的服务消费者。如果此时 A 发生故障，则会引起 B 的不可用，而 B 的不可用又将导致 C 和 D 的不可用，当这种不可用像滚雪球一样逐渐放大的时候，雪崩效应就形成了。

## 16.2 熔断器（CircuitBreaker）

熔断器的原理很简单，如同电力过载保护器。它可以实现快速失败，如果它在一段时间内侦测到许多类似的错误，就会强迫其以后的多个调用快速失败，不再访问远程服务器，从而防止应用程序不断地尝试执行可能会失败的操作，使得应用程序继续执行而不用等待修正错误，或者浪费 CPU 时间去等到长时间的超时产生。熔断器也可以使应用程序能够诊断错误是否已经修正，如果已经修正，应用程序会再次尝试调用操作。熔断器模式就像是那些容易导致错误操作的一种代理。这种代理能够记录最近调用发生错误的次数，然后决定使用允许操作继续，或者立即返回错误。熔断器是保护服务高可用的最后一道防线。

## 16.3 Hystrix 特性

### 16.3.1 断路器机制

断路器很好理解，当 Hystrix Command 请求后端服务失败数量超过一定比例（默认为

50%），断路器会切换到开路状态（Open）。这时所有请求会直接失败而不会发送到后端服务。断路器保持在开路状态一段时间后（默认为 5 秒），自动切换到半开路状态（HALF-OPEN）。这时会判断下一次请求的返回情况，如果请求成功，断路器切回闭路状态（CLOSED），否则重新切换到开路状态（OPEN）。Hystrix 的断路器就像我们家庭电路中的保险丝，一旦后端服务不可用，断路器就会直接切断请求链，避免发送大量无效请求，从而影响系统吞吐量，并且断路器有自我检测并恢复的能力。

### 16.3.2 fallback

fallback 相当于降级操作。对于查询操作，我们可以实现一个 fallback 方法，当请求后端服务出现异常的时候，可以使用 fallback 方法返回的值。fallback 方法的返回值一般是设置的默认值或者来自缓存。

### 16.3.3 资源隔离

在 Hystrix 中，主要通过线程池来实现资源隔离。通常在使用的时候我们会根据调用的远程服务划分出多个线程池。例如，调用产品服务的 Command 放入 A 线程池，调用账户服务的 Command 放入 B 线程池。这样做的主要优点是运行环境被隔离开了。这样就算调用服务的代码存在 bug 或者由于其他原因导致自己所在线程池被耗尽，也不会对系统的其他服务造成影响，但是带来的代价就是维护多个线程池会对系统带来额外的性能开销。如果是对性能有严格要求而且确信自己调用服务的客户端代码不会出问题，就可以使用 Hystrix 的信号模式（Semaphores）来隔离资源。

## 16.4 Feign Hystrix

因为 Feign 中已经依赖了 Hystrix，所以在 Maven 配置上不用做任何改动就可以使用了，我们可以在 mango-consumer 项目中直接改造。

### 16.4.1 修改配置

在配置文件中添加配置，开启 Hystrix 熔断器。

application.yml

```
#开启熔断器
feign:
  hystrix:
    enabled: true
```

### 16.4.2 创建回调类

创建一个回调类 MangoProducerHystrix，实现 MangoProducerService 接口，并实现对应的方法，返回调用失败后的信息。

MangoProducerHystrix.java

```
@Component
public class MangoProducerHystrix implements MangoProducerService {

    @RequestMapping("/hello")
    public String hello() {
    return "sorry, hello service call failed.";
}

}
```

添加 fallback 属性。修改 MangoProducerService，在 @FeignClient 注解中加入 fallback 属性，绑定我们创建的失败回调处理类。

MangoProducerService.java

```
@FeignClient(name = "mango-producer", fallback = MangoProducerHystrix.class)
public interface MangoProducerService {

    @RequestMapping("/hello")
    public String hello();

}
```

到此，所有改动代码就完成了，就是这么简单。

### 16.4.3 页面测试

启动成功之后，多次访问 http://localhost:8005/feign/call，结果如同之前一样交替返回"hello Mango！"和"hello Mango2！"，说明熔断器的启动不会影响正常服务的访问，如图 16-1 所示。

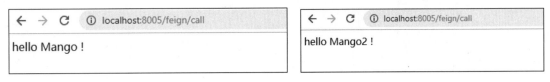

图 16-1

把 mango-producer 服务停掉，再次访问，返回我们提供的熔断回调信息，熔断成功，mango-producer2 服务正常，如图 16-2 所示。

重启 mango-producer 服务，再次访问，发现服务又可以访问了，说明熔断器具有自我诊断修复的功能，如图 16-3 所示。

图 16-2

图 16-3

> 在重启成功之后，可能需要一些时间，等待熔断器进行自我诊断和修复完成之后，方可正常提供服务。

# 16.5 Hystrix Dashboard

Hystrix-dashboard 是一款针对 Hystrix 进行实时监控的工具，通过 Hystrix Dashboard 我们可以直观地看到各 Hystrix Command 的请求响应时间、请求成功率等数据。

## 16.5.1 添加依赖

新建一个 mango-hystrix 工程，修改 pom 文件，添加相关依赖。

pom.xml

```xml
<!-- spring boot -->
<dependency>
    <groupId>org.springframework.boot</groupId>
    <artifactId>spring-boot-starter</artifactId>
</dependency>
<!--consul-->
<dependency>
    <groupId>org.springframework.cloud</groupId>
    <artifactId>spring-cloud-starter-consul-discovery</artifactId>
</dependency>
<!--actuator-->
<dependency>
    <groupId>org.springframework.boot</groupId>
    <artifactId>spring-boot-starter-actuator</artifactId>
</dependency>
<!--spring-boot-admin-->
<dependency>
```

```xml
    <groupId>de.codecentric</groupId>
    <artifactId>spring-boot-admin-starter-client</artifactId>
    <version>${spring.boot.admin.version}</version>
</dependency>
<!--hystrix-dashboard-->
<dependency>
    <groupId>org.springframework.cloud</groupId>
    <artifactId>spring-cloud-starter-netflix-hystrix-dashboard</artifactId>
</dependency>
```

Spring Cloud 依赖

```xml
<dependencyManagement>
    <dependencies>
        <dependency>
            <groupId>org.springframework.cloud</groupId>
            <artifactId>spring-cloud-dependencies</artifactId>
            <version>${spring-cloud.version}</version>
            <type>pom</type>
            <scope>import</scope>
        </dependency>
    </dependencies>
</dependencyManagement>
```

### 16.5.2 启动类

在启动类中添加注解 @EnableHystrixDashboard，开启熔断监控支持。

MangoHystrixApplication.java

```java
@EnableHystrixDashboard
@EnableDiscoveryClient
@SpringBootApplication
public class MangoHystrixApplication {

    public static void main(String[] args) {
        SpringApplication.run(MangoHystrixApplication.class, args);
    }

}
```

### 16.5.3 自定义 Banner

在 resources 目录下新建一个自定义 banner 文件。

**banner.txt**

```
.__                   __        .__                
|  |__ ___.__._____/  |_____|__|__ ___         
|  |  <   |  |/  ___/\   __\_  __ \  |  |  \       
|   Y  \___  |\___ \  |  |  |  | \/  |  |  /       
|___|  / ____/____  > |__|  |__|  |__|_/\_ \       
     \/\/         \/                       \/      

===== Mango Application == Version: 1.0 =====
```

### 16.5.4 配置文件

修改配置文件，把服务注册到注册中心。

**application.yml**

```yml
server:
  port: 8501
spring:
  application:
    name: mango-hystrix
  cloud:
    consul:
      host: localhost
      port: 8500
      discovery:
        serviceName: ${spring.application.name} # 注册到 consul 的服务名称
```

### 16.5.5 配置监控路径

注意，如果使用的是 2.x 等比较新的版本，就需要在 Hystrix 的消费端配置监控路径。打开消费端 mango-consumer 工程，添加依赖。

**pom.xml**

```xml
<!--actuator-->
<dependency>
    <groupId>org.springframework.boot</groupId>
    <artifactId>spring-boot-starter-actuator</artifactId>
</dependency>
<!--hystrix-dashboard-->
<dependency>
    <groupId>org.springframework.cloud</groupId>
    <artifactId>spring-cloud-starter-netflix-hystrix-dashboard</artifactId>
</dependency>
```

修改启动类，添加服务监控路径配置。

MangoConsumerApplication.java

```
// 此配置是为了服务监控而配置，与服务容错本身无关，
// ServletRegistrationBean 因为 springboot 的默认路径不是"/hystrix.stream"，
// 只要在自己的项目里配置上下面的 servlet 就可以了
@Bean
public ServletRegistrationBean getServlet() {
    HystrixMetricsStreamServlet streamServlet = new HystrixMetricsStreamServlet();
    ServletRegistrationBean registrationBean = new ServletRegistrationBean(streamServlet);
    registrationBean.setLoadOnStartup(1);
    registrationBean.addUrlMappings("/hystrix.stream");
    registrationBean.setName("HystrixMetricsStreamServlet");
    return registrationBean;
}
```

### 16.5.6　页面测试

先后启动 monitor、producer、consumer、hystrix 服务，访问 http://localhost:8501/hystrix，会看到如图 16-4 所示的界面。

图 16-4

此时没有任何具体的监控信息，需要输入要监控的消费者地址及监控信息的轮询时间和标题。

Hystrix Dashboard 共支持以下 3 种不同的监控方式：

（1）单体 Hystrix 消费者：通过 URL http://hystrix-app:port/hystrix.stream 开启，实现对具体某个服务实例的监控。

（2）默认集群监控：通过 URL http://turbine-hostname:port/turbine.stream 开启，实现对默认集群的监控。

(3)自定集群监控：通过 URL http://turbine-hostname:port/turbine.stream?cluster=[clusterName] 开启，实现对 clusterName 集群的监控。

这里先介绍对单体 Hystrix 消费者的监控，后面整合 Turbine 集群的时候再说明后两种监控方式。

首先，访问 http://localhost:8005/feign/call，查看要监控的服务是否可以正常访问，如图 16-5 所示。

图 16-5

确认服务可以正常访问之后，在监控地址内输入 http://localhost:8005/hystrix.stream，然后单击 Monitor Stream 开始监控，如图 16-6 所示。

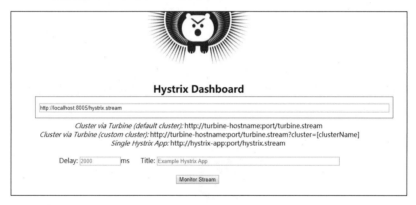

图 16-6

刚进去，页面先显示 loading... 信息，在多次间断访问 http://localhost:8005/feign/call 之后，统计图表信息如图 16-7 所示。

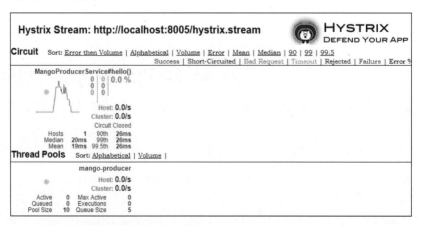

图 16-7

各项指标的含义如图 16-8 所示。

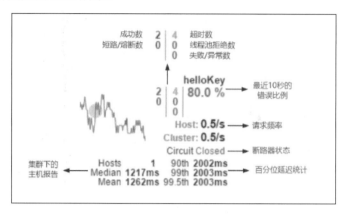

图 16-8

# 16.6 Spring Cloud Turbine

上面我们集成了 Hystrix Dashboard，使用 Hystrix Dashboard 可以看到单个应用内的服务信息。显然这是不够的，我们还需要一个工具能让我们汇总系统内多个服务的数据并显示到 Hystrix Dashboard 上，这个工具就是 Turbine。

## 16.6.1 添加依赖

修改 mango-hystrix 的 pom 文件，添加 turbine 依赖包。

注意

因为我们使用的注册中心是 Consul，所以需要排除默认的 euraka 包，不然会出现冲突，导致启动过程出错。

pom.xml

```
<!--turbine-->
<dependency>
    <groupId>org.springframework.cloud</groupId>
    <artifactId>spring-cloud-starter-netflix-turbine</artifactId>
    <exclusions>
        <exclusion>
            <groupId>org.springframework.cloud</groupId>
            <artifactId>spring-cloud-starter-netflix-eureka-client</artifactId>
        </exclusion>
    </exclusions>
</dependency>
```

## 16.6.2 启动类

为启动类添加@EnableTurbine 注解，开启 Turbine 支持。

MangoHystrixApplication.java

```
@EnableTurbine
@EnableHystrixDashboard
@EnableDiscoveryClient
@SpringBootApplication
public class MangoHystrixApplication {

    public static void main(String[] args) {
        SpringApplication.run(MangoHystrixApplication.class, args);
    }

}
```

## 16.6.3 配置文件

修改配置文件，添加 Turbine 的配置信息。

application.yml

```
server:
  port: 8501
spring:
  application:
    name: mango-hystrix
  cloud:
    consul:
      host: localhost
      port: 8500
      discovery:
        serviceName: ${spring.application.name}     # 注册到 consul 的服务名称
turbine:
  instanceUrlSuffix: hystrix.stream                 # 指定收集路径
  appConfig: kitty-consumer         # 指定了需要收集监控信息的服务名,多个以","进行区分
  clusterNameExpression: "'default'"
    # 指定集群名称,若为 default 则为默认集群,多个集群则通过此配置区分
  combine-host-port: true
    # 默认为 false,服务以 host 进行区分,若设置为 true 则以 host+port 进行区分
```

## 16.6.4 测试效果

依次启动 monitor、producer、consumer、hystrix 服务，确认服务启动无误之后访问 http://localhost:8501/hystrix，输入 http://localhost:8501/turbine.stream，查看熔断监控图表信息，如图 16-9 所示。

图 16-9

图 16-10 所示就是利用 Turbine 聚合多个 Hytrix 消费者的熔断监控信息结果。内存允许的话，可以多启动几个消费者查看。

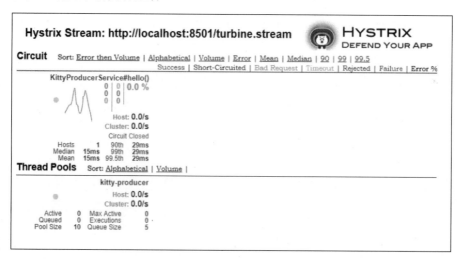

图 16-10

# 第 17 章 服务网关（Zuul）

## 17.1 技术背景

前面我们通过 Ribbon 或 Feign 实现了微服务之间的调用和负载均衡，那我们的各种微服务又要如何提供给外部应用调用呢？

因为是 REST API 接口，所以外部客户端直接调用各个微服务是没有问题的，但是出于种种原因，这并不是一个好的选择。

让客户端直接与各个微服务通信，会有以下几个问题：

- 客户端会多次请求不同的微服务，增加客户端的复杂性。
- 存在跨域请求，在一定场景下处理会变得相对比较复杂。
- 实现认证复杂，每个微服务都需要独立认证。
- 难以重构，项目迭代可能导致微服务重新划分。如果客户端直接与微服务通信，那么重构将会很难实施。
- 如果某些微服务使用了防火墙/浏览器不友好的协议，直接访问会有一定困难。
- 面对类似上面的问题，我们要如何解决呢？答案就是：服务网关！

使用服务网关具有以下几个优点：

- 易于监控。可在微服务网关收集监控数据并将其推送到外部系统进行分析。
- 易于认证。可在服务网关上进行认证，然后转发请求到微服务，无须在每个微服务中进行认证。
- 客户端只跟服务网关打交道，减少了客户端与各个微服务之间的交互次数。
- 多渠道支持，可以根据不同客户端（Web端、移动端、桌面端等）提供不同的 API 服务网关。

## 17.2 Spring Cloud Zuul

服务网关是微服务架构中一个不可或缺的部分。在通过服务网关统一向外系统提供 REST

API 的过程中，除了具备服务路由、均衡负载功能之外，它还具备了权限控制等功能。

Spring Cloud Netflix 中的 Zuul 就担任了这样的一个角色，为微服务架构提供了前门保护的作用，同时将权限控制这些较重的非业务逻辑内容迁移到服务路由层面，使得服务集群主体能够具备更高的可复用性和可测试性。

在 Spring Cloud 体系中，Spring Cloud Zuul 封装了 Zuul 组件，作为一个 API 网关，负责提供负载均衡、反向代理和权限认证。

## 17.3 Zuul 工作机制

### 17.3.1 过滤器机制

Zuul 的核心是一系列的 filters，其作用类似 Servlet 框架的 Filter，Zuul 把客户端请求路由到业务处理逻辑的过程中，这些 filter 在路由的特定时期参与了一些过滤处理，比如实现鉴权、流量转发、请求统计等功能。Zuul 的整个运行机制可以用图 17-1 来描述。

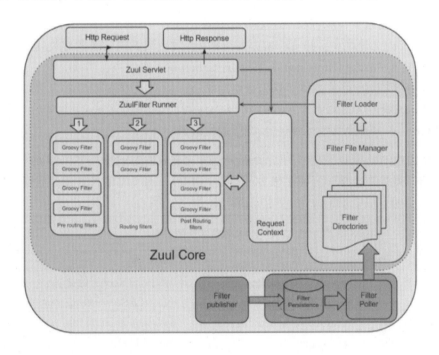

图 17-1

### 17.3.2 过滤器的生命周期

Filter 的生命周期有 4 个，分别是"PRE""ROUTING""POST""ERROR"，整个生命周期可以用图 17-2 来表示。

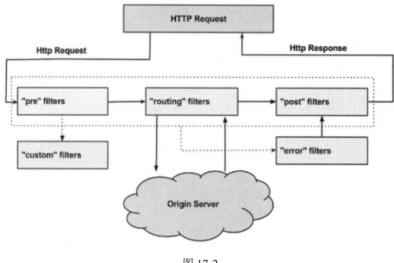

图 17-2

基于 Zuul 的这些过滤器可以实现各种丰富的功能，而这些过滤器类型则对应于请求的典型生命周期。

- PRE：这种过滤器在请求被路由之前调用。我们可利用这种过滤器实现身份验证、在集群中选择请求的微服务、记录调试信息等。
- ROUTING：这种过滤器将请求路由到微服务。这种过滤器用于构建发送给微服务的请求，并使用 Apache HttpClient 或 Netfilx Ribbon 请求微服务。
- POST：这种过滤器在路由到微服务以后执行。这种过滤器可用来为响应添加标准的 HTTP Header、收集统计信息和指标、将响应从微服务发送给客户端等。
- ERROR：在其他阶段发生错误时执行该过滤器。

除了默认的过滤器类型，Zuul 还允许我们创建自定义的过滤器类型。例如，我们可以定制一种 STATIC 类型的过滤器，直接在 Zuul 中生成响应，而不将请求转发到后端的微服务。

Zuul 默认实现了很多 Filter，如表 17-1 所示。

表 17-1　Zuul 默认实现的 Filter

| 类 型 | 顺 序 | 过 滤 器 | 功 能 |
| --- | --- | --- | --- |
| pre | -3 | ServletDetectionFilter | 标记处理 Servlet 的类型 |
| pre | -2 | Servlet30WrapperFilter | 包装 HttpServletRequest 请求 |
| pre | -1 | FormBodyWrapperFilter | 包装请求体 |
| route | 1 | DebugFilter | 标记调试标志 |
| route | 5 | PreDecorationFilter | 处理请求上下文，供后续使用 |
| route | 10 | RibbonRoutingFilter | serviceId 请求转发 |
| route | 100 | SimpleHostRoutingFilter | url 请求转发 |
| route | 500 | SendForwardFilter | forward 请求转发 |
| post | 0 | SendErrorFilter | 处理有错误的请求响应 |
| post | 1000 | SendResponseFilter | 处理正常的请求响应 |

### 17.3.3 禁用指定的 Filter

可以在 application.yml 中配置需要禁用的 filter，格式为 zuul.<SimpleClassName>.<filterType>.disable=true。比如要禁用 org.springframework.cloud.netflix.zuul.filters.post.SendResponseFilter，进行如下设置即可：

```yaml
zuul:
  SendResponseFilter:
    post:
      disable: true
```

自定义 Filter。实现自定义滤器需要继承 ZuulFilter，并实现 ZuulFilter 中的抽象方法。

```java
public class MyFilter extends ZuulFilter {
    @Override
    String filterType() {
        return "pre"; // 定义filter的类型，有pre、route、post、error四种
    }

    @Override
    int filterOrder() {
        return 5; // 定义filter的顺序，数字越小，表示顺序越高，越先执行
    }

    @Override
    boolean shouldFilter() {
        return true; // 表示是否需要执行该filter，true表示执行，false表示不执行
    }

    @Override
    Object run() {
        return null; // filter需要执行的具体操作
    }
}
```

## 17.4 实现案例

### 17.4.1 新建工程

新建一个项目 mango-zuul 作为服务网关，工程结构如图 17-3 所示。

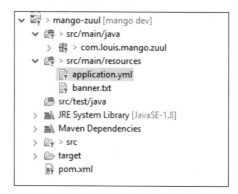

图 17-3

### 17.4.2 添加依赖

添加 consul、zuul 的相关依赖。

pom.xml

```xml
<dependencies>
    <dependency>
        <groupId>org.springframework.cloud</groupId>
        <artifactId>spring-cloud-starter-netflix-zuul</artifactId>
    </dependency>
    <dependency>
        <groupId>org.springframework.cloud</groupId>
        <artifactId>spring-cloud-starter-consul-discovery</artifactId>
    </dependency>
</dependencies>

<dependencyManagement>
    <dependencies>
        <dependency>
            <groupId>org.springframework.cloud</groupId>
            <artifactId>spring-cloud-dependencies</artifactId>
            <version>${spring-cloud.version}</version>
            <type>pom</type>
            <scope>import</scope>
        </dependency>
    </dependencies>
</dependencyManagement>
```

### 17.4.3 启动类

为启动类添加 @EnableZuulProxy 注解，开启服务网关支持。

MangoZuulApplication.java

```java
@EnableZuulProxy
@SpringBootApplication
public class MangoZuulApplication {

    public static void main(String[] args) {
        SpringApplication.run(MangoZuulApplication.class, args);
    }

}
```

### 17.4.4 配置文件

配置启动端口为 8010，注册服务到注册中心，配置 Zuul 转发规则。这里配置在访问 locathost:8010/feign/call 和 ribbon/call 时调用消费者对应接口。

application.yml

```yml
server:
  port: 8010
spring:
  application:
    name: mango-zuul
  cloud:
    consul:
      host: localhost
      port: 8500
      discovery:
        serviceName: ${spring.application.name}    # 注册到 consul 的服务名称
zuul:
  routes:
    ribbon:
      path: /ribbon/**
      serviceId: mango-consumer   # 转发到消费者 /ribbon/
    feign:
      path: /feign/**
      serviceId: mango-consumer   # 转发到消费者 /feign/
```

### 17.4.5 页面测试

依次启动注册中心、监控、服务提供者、服务消费者、服务网关等项目。

访问 http://localhost:8010/ribbon/call，效果如图 17-4 所示。

访问 http://localhost:8010/feign/call，效果如图 17-5 所示。

图 17-4   图 17-5

说明 Zuul 已经成功转发请求，并成功调用后端微服务。

### 17.4.6 配置接口前缀

如果想给每个服务的 API 接口加上一个前缀，可使用 zuul.prefix 进行配置。例如，http://localhost:8010/v1/feign/call，即在所有的 API 接口上加一个 v1 作为版本号。

```
zuul:
  prefix: /v1
  routes:
    ribbon:
      path: /ribbon/**
      serviceId: kitty-consumer    # 转发到消费者 /ribbon/
    feign:
      path: /feign/**
      serviceId: kitty-consumer    # 转发到消费者 /feign/
```

### 17.4.7 默认路由规则

上面我们是通过添加路由配置进行请求转发的，内容如下：

```
zuul:
  routes:
    ribbon:
      path: /ribbon/**
      serviceId: kitty-consumer    # 转发到消费者 /ribbon/
    feign:
      path: /feign/**
      serviceId: kitty-consumer    # 转发到消费者 /feign/
```

但是如果后端微服务非常多，每一个都这样配置还是挺麻烦的。Spring Cloud Zuul 已经帮我们做了默认配置。默认情况下，Zuul 会代理所有注册到注册中心的微服务，并且 Zuul 的默认路由规则如下：http://ZUUL_HOST:ZUUL_PORT/微服务在注册中心的 serviceId/** 会被转发到 serviceId 对应的微服务。如果遵循默认路由规则，基本上就没什么配置了。

比如我们直接通过 serviceId/feign/call 的方式访问，也是可以正常访问的。访问 http://localhost:8010/mango-consumer/feign/call，结果也是一样的，如图 17-6 所示。

图 17-6

## 17.4.8 路由熔断

Zuul 作为 Netflix 组件，可以与 Ribbon、Eureka 和 Hystrix 等组件相结合，实现负载均衡、熔断器的功能。默认情况下 Zuul 和 Ribbon 相结合，实现了负载均衡。实现熔断器功能需要实现 FallbackProvider 接口。实现该接口有两个方法：一个是 getRoute()，用于指定熔断器功能应用于哪些路由的服务；另一个方法是 fallbackResponse()，为进入熔断器功能时执行的逻辑。

创建 MyFallbackProvider 类，getRoute()方法返回"mango-consumer"，只针对 consumer 服务进行熔断。如果需要所有的路由服务都加熔断功能，需要在 getRoute()方法上返回" * "匹配符。getBody()方法返回发送熔断时的反馈信息，这里在发送熔断时返回信息"Sorry, the service is unavailable now."。

MyFallbackProvider.java

```
@Component
public class MyFallbackProvider implements FallbackProvider {
    @Override
    public String getRoute() {
        return "mango-consumer";
    }

    @Override
    public ClientHttpResponse fallbackResponse(String route, Throwable cause) {
        System.out.println("route:"+route);
        System.out.println("exception:"+cause.getMessage());
        return new ClientHttpResponse() {
            @Override
            public HttpStatus getStatusCode() throws IOException {
                return HttpStatus.OK;
            }
            @Override
            public int getRawStatusCode() throws IOException {
                return 200;
            }
            @Override
            public String getStatusText() throws IOException {
                return "ok";
            }
```

```
            @Override
            public void close() { }
            @Override
            public InputStream getBody() throws IOException {
                return new ByteArrayInputStream("Sorry, the service is unavailable now.".getBytes());
            }
            @Override
            public HttpHeaders getHeaders() {
                HttpHeaders headers = new HttpHeaders();
                headers.setContentType(MediaType.APPLICATION_JSON);
                return headers;
            }
        };
    }
}
```

重新启动，访问 http://localhost:8010/mango-consumer/feign/call，可以正常访问，如图 17-7 所示。

停掉 mango-consumer 服务，访问 http://localhost:8010/mango-consumer/feign/call，返回效果如图 17-8 所示。

图 17-7　　　　　　　　　　　　　　　图 17-8

结果返回了我们自定义的信息，说明我们自定义的熔断器已经起作用了。

### 17.4.9　自定义 Filter

创建一个 MyFilter，继承 ZuulFilter 类，覆写 run()方法逻辑，在转发请求前进行 token 认证，如果请求没有携带 token，返回"there is no request token"提示。

**MyFilter.java**

```
@Component
public class MyFilter extends ZuulFilter {
    private static Logger log=LoggerFactory.getLogger(MyFilter.class);
    @Override
    public String filterType() {
        return "pre"; // 定义 filter 的类型，有 pre、route、post、error 四种
    }
    @Override
```

```java
public int filterOrder() {
    return 0; // 定义 filter 的顺序，数字越小，表示顺序越高、越先执行
}
@Override
public boolean shouldFilter() {
    return true; // 表示是否需要执行该 filter，true 表示执行，false 表示不执行
}
@Override
public Object run() throws ZuulException {
    // filter 需要执行的具体操作
    RequestContext ctx = RequestContext.getCurrentContext();
    HttpServletRequest request = ctx.getRequest();
    String token = request.getParameter("token");
    System.out.println(token);
    if(token==null){
        ctx.setSendZuulResponse(false);
        ctx.setResponseStatusCode(401);
        try {
            ctx.getResponse().getWriter().write("there is no request token");
        } catch (IOException e) {
            e.printStackTrace();
        }
        return null;
    }
    return null;
}
}
```

这样，Zuul 就会自动加载 Filter 执行过滤了。重新启动 Zuul 项目，访问 http://localhost:8010/mango-consumer/feign/call，结果如图 17-9 所示。

请求时带上 token，访问 http://localhost:8010/mango-consumer/feign/call?token=111，接口返回正确结果，如图 17-10 所示。

图 17-9                                    图 17-10

Zuul 作为 API 服务网关，不同的客户端使用不同的负载将请求统一分发到后端的 Zuul，再由 Zuul 转发到后端服务。为了保证 Zuul 的高可用性，前端可以同时开启多个 Zuul 实例进行负载均衡。另外，在 Zuul 的前端还可以使用 Nginx 或者 F5 再次进行负载转发，从而保证 Zuul 的高可用性。

# 第 18 章 链路追踪（Sleuth、ZipKin）

## 18.1 技术背景

在微服务架构中，随着业务发展，系统拆分导致系统调用链路愈发复杂，一个看似简单的前端请求可能最终需要调用很多次后端服务才能完成，那么当整个请求出现问题时，我们很难得知到底是哪个服务出了问题导致的,这时就需要解决一个问题，即如何快速定位服务故障点，分布式系统调用链追踪技术就此诞生了。

## 18.2 ZipKin

ZipKin 是一个由 Twitter 公司提供并开放源代码分布式的跟踪系统，它可以帮助收集服务的时间数据，以解决微服务架构中的延迟问题，包括数据的收集、存储、查找和展现。

每个服务向 ZipKin 报告定时数据，ZipKin 会根据调用关系通过 ZipKin UI 生成依赖关系图，展示多少跟踪请求经过了哪些服务，该系统让开发者可通过一个 Web 前端轻松地收集和分析数据，例如用户每次请求服务的处理时间等，可非常方便地监测系统中存在的瓶颈。

ZipKin 提供了可插拔数据存储方式：In-Memory、MySQL、Cassandra 以及 Elasticsearch。我们可以根据需求选择不同的存储方式，生成环境一般都需要持久化。我们这里采用 Elasticsearch 作为 ZipKin 的数据存储器。

## 18.3 Spring Cloud Sleuth

一般而言，一个分布式服务追踪系统，主要由 3 部分组成：数据收集、数据存储和数据展示。

Spring Cloud Sleuth 为服务之间的调用提供链路追踪，通过 Sleuth 可以很清楚地了解到一个服务请求经过了哪些服务，每个服务处理花费了多长时间，从而让我们可以很方便地理清各微服务间的调用关系。此外，Sleuth 还可以帮助我们：

- 耗时分析：通过 Sleuth 可以很方便地了解到每个采样请求的耗时，从而分析出哪些服务调用比较耗时。
- 可视化错误：对于程序未捕捉的异常，可以通过集成 ZipKin 服务在界面上看到。
- 链路优化：对于调用比较频繁的服务，可以针对这些服务实施一些优化措施。

Spring Cloud Sleuth 可以结合 ZipKin，将信息发送到 ZipKin，利用 ZipKin 的存储来存储信息，利用 ZipKin UI 来展示数据。

# 18.4 实现案例

在早前的 Spring Cloud 版本里是需要自建 ZipKin 服务端的，但是从 Spring Cloud 2.0 以后，官方已经不支持自建 Server 了，改成提供编译好的 jar 包供用户使用。这里使用的是 2.0 以后的版本，自建 Server 的方式请自行到百度上查看。这里我们使用 Docker 方式部署 ZipKin 服务，并采用 Elasticsearch 作为 ZipKin 的数据存储器。

## 18.4.1 下载镜像

此前请先安装好 Docker 环境，使用以下命令分别拉取 ZipKin 和 Elasticsearch 镜像。

```
docker pull openzipkin/zipkin
docker pull docker.elastic.co/elasticsearch/elasticsearch:6.3.0
```

通过 docker images 查看下载镜像，如图 18-1 所示。

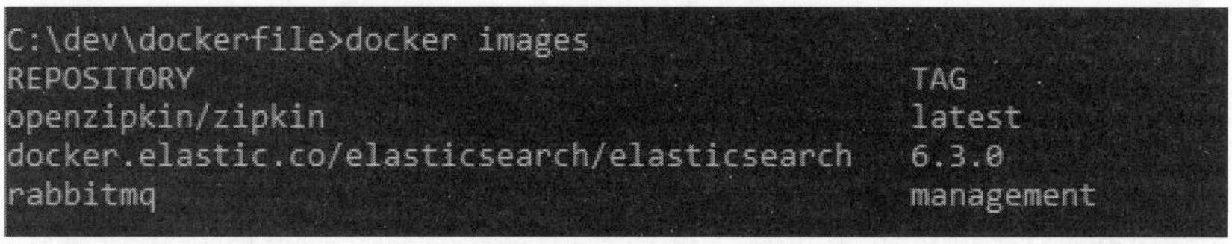

图 18-1

## 18.4.2 编写启动文件

在本地创建如下文件夹结构，其中 data 目录用来存放 Elasticsearch 存储的数据。

```
dockerfile
   |- elasticsearch
   |    |- data
   |- docker-compose.yml
```

编写 docker-compose 文件，主要作用是批量启动容器，避免在使用多个容器的时候逐个启动。

docker-compose.yml

```yaml
version: "3"
services:

 elasticsearch:
   image: docker.elastic.co/elasticsearch/elasticsearch:6.3.0
   container_name: elasticsearch
   restart: always
   networks:
     - elk
   ports:
     - "9200:9200"
     - "9300:9300"
   volumes:
     - ../elasticsearch/data:/usr/share/elasticsearch/data

 zipkin:
   image: openzipkin/zipkin:latest
   container_name: zipkin
   restart: always
   networks:
     - elk
   ports:
     - "9411:9411"
   environment:
     - STORAGE_TYPE=elasticsearch
     - ES_HOSTS=elasticsearch
networks:
   elk:
```

关于 docker-compose.yml 文件的格式及相关内容，请自行到百度网站搜索。

### 18.4.3 启动服务

以命令模式进入 dockerfile 目录，执行启动的命令如下：

```
docker-compose up -d
```

执行过程如图 18-2 所示。

```
C:\dev\dockerfile>docker-compose up -d
Creating zipkin ...
Creating elasticsearch ...
Creating zipkin
Creating elasticsearch ... done
```

图 18-2

## 第 18 章 链路追踪（Sleuth、ZipKin）

执行完成之后，通过 docker ps 命令查看，发现 ZipKin 和 Elasticsearch 确实启动起来了，如图 18-3 所示。

图 18-3

到这里，ZipKin 服务端就搭建起来了，访问 http://localhost:9411，效果如图 18-4 所示。因为还没有客户端，所以还没有数据。

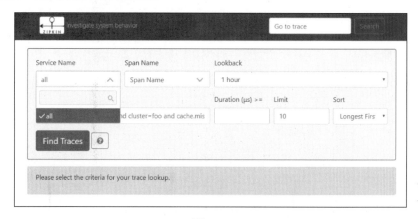

图 18-4

注意

这里我们采用了 Elasticsearch 作为存储方式，如果想简单通过内存方式启动，无须安装 Elasticsearch，直接启动一个 ZipKin 容器即可。

docker run -d -p 9411:9411 openzipkin/zipkin

如果想使用其他方式（如数据库等）存储，请查询相关配置文档。

ZipKin 服务端已经搭建完成了，接下来我们来实现客户端。

### 18.4.4 添加依赖

修改 mango-consumer 项目 Maven 配置，添加 ZipKin 依赖。

pom.xml

```
<!-- zipkin -->
<dependency>
```

177

```
        <groupId>org.springframework.cloud</groupId>
        <artifactId>spring-cloud-starter-zipkin</artifactId>
</dependency>
```

### 18.4.5 配置文件

修改配置文件，添加如下 ZipKin 配置。

**application.yml**

```
spring:
 zipkin:
  base-url: http://localhost:9411/
 sleuth:
  sampler:
   probability: 1    #样本采集量，默认为0.1
```

### 18.4.6 页面测试

先后启动注册中心、服务监控、服务提供者、服务消费者。反复访问几次 http://localhost:8005/feign/call，产生 ZipKin 数据，如图 18-5 所示。

图 18-5

再次访问 http://localhost:9411，发现出现了我们刚刚访问的服务，选择服务并单击 Find Traces 按钮追踪，如图 18-6 所示。

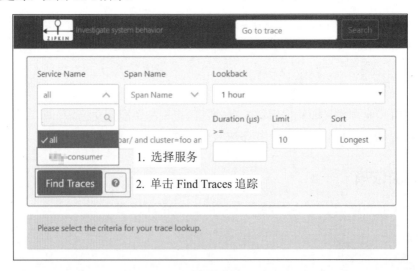

图 18-6

追踪之后，页面显示出相关的服务调用信息，如图 18-7 所示。

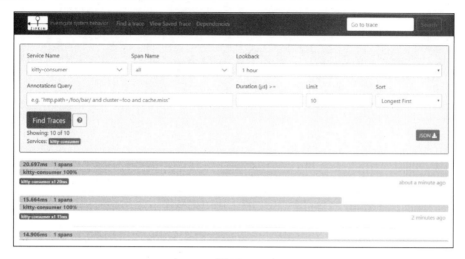

图 18-7

单击调用记录查看详情页面，可以看到每一个服务所耗费的时间和顺序，如图 18-8 所示。

图 18-8

# 第 19 章 配置中心（Config、Bus）

## 19.1 技术背景

如今微服务架构盛行，在分布式系统中，项目日益庞大，子项目日益增多，每个项目都散落着各种配置文件，且随着服务的增加而不断增多。此时，往往某一个基础服务信息变更都会导致一系列服务的更新和重启，运维也是苦不堪言，而且还很容易出错。配置中心便由此应运而生了。

目前市面上开源的配置中心很多，像 Spring 家族的 Spring Cloud Config、Apache 的 Apache Commons Configuration、淘宝的 diamond、百度的 disconf、360 的 QConf 等，都是为了解决这类问题。当下 Spring 体系大行其道，我们也优先选择了 Spring Cloud Config。

## 19.2 Spring Cloud Config

Spring Cloud Config 是一套为分布式系统中的基础设施和微服务应用提供集中化配置的管理方案，分为服务端与客户端两个部分。服务端也称为分布式配置中心，是一个独立的微服务应用，用来连接配置仓库并为客户端提供获取配置信息。客户端是微服务架构中的各个微服务应用或基础设施，它们通过指定的配置中心来管理服务相关的配置内容，并在启动的时候从配置中心获取和加载配置信息。

Spring Cloud Config 对服务端和客户端中的环境变量和属性配置实现了抽象映射，所以除了适用于 Spring 应用，也是可以在任何其他语言应用中使用的。Spring Cloud Config 实现的配置中心默认采用 Git 来存储配置信息，所以使用 Spring Cloud Config 构建的配置服务器天然就支持对微服务应用配置信息的版本管理，并且可以通过 Git 客户端工具非常方便地管理和访问配置内容。当然它也提供了对其他存储方式的支持，比如 SVN 仓库、本地化文件系统等。

## 19.3 实现案例

### 19.3.1 准备配置文件

首先在 GIT 下新建一个 config-repo 目录，用来存放配置文件，如图 19-1 所示。这里分别模拟了 3 个环境的配置文件。

图 19-1

分别编辑 3 个文件，配置 hello 属性的值为 "consumer.hello=hello, xx configurations."，如图 19-2 所示。

图 19-2

### 19.3.2 服务端实现

#### 1. 新建工程

新建 mango-config 工程，作为配置中心的服务端，负责把 GIT 仓库的配置文件发布为 RESTFul 接口，如图 19-3 所示。

#### 2. 添加依赖

除了 Spring Cloud 依赖之外，另需添加配置中心依赖包。

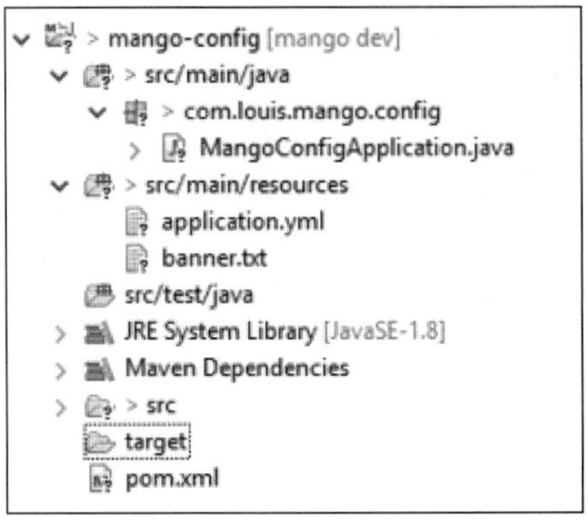

```
mango-config [mango dev]
  src/main/java
    com.louis.mango.config
      MangoConfigApplication.java
  src/main/resources
    application.yml
    banner.txt
  src/test/java
  JRE System Library [JavaSE-1.8]
  Maven Dependencies
  src
  target
  pom.xml
```

图 19-3

pom.xml

```xml
<!--spring config-->
<dependencies>
    <dependency>
        <groupId>org.springframework.cloud</groupId>
        <artifactId>spring-cloud-config-server</artifactId>
    </dependency>
</dependencies>
```

### 3. 启动类

启动类添加注解@EnableConfigServer，开启配置服务支持。

MangoConfigApplication.java

```java
@EnableDiscoveryClient
@EnableConfigServer
@SpringBootApplication
public class MangoConfigApplication {

    public static void main(String[] args) {
        SpringApplication.run(MangoConfigApplication.class, args);
    }

}
```

### 4. 配置文件

修改配置文件，添加如下内容。如果是私有仓库，需要填写用户名、密码；如果是公开仓库，可以不配置密码。

application.yml

```yaml
server:
  port: 8020
spring:
  application:
    name: mango-config
  cloud:
    consul:
      host: localhost
      port: 8500
      discovery:
        serviceName: ${spring.application.name}     # 注册到 consul 的服务名称
    config:
      label: master  # git 仓库分支
      server:
        git:
          uri: https://gitee.com/liuge1988/mango.git  # 配置 git 仓库的地址
          search-paths: src/config-repo  # git 仓库地址下的相对地址,可配置多个,用,分割
          username: username   # git 仓库的账号
          password: password   # git 仓库的密码
```

Spring Cloud Config 也提供本地存储配置方式,只需设置属性 spring.profiles.active=native,Config Server 会默认从应用的 src/main/resource 目录下检索配置文件。另外,也可以通过 spring.cloud.config.server.native.searchLocations=file:D:/properties/属性来指定配置文件的位置。虽然 Spring Cloud Config 提供了这样的功能,但是为了更好地支持内容管理和版本控制,还是推荐使用 GIT 的方式。

5. 页面测试

启动注册中心,配置中心服务,访问 http://localhost:8020/ consumer/dev,返回 dev 配置文件的信息。

```
{
    "name": "consumer",
    "profiles": ["dev"],
    "label": null,
    "version": "feab08409500b3626e3f70f3b4d24f7d0cb93efc",
    "state": null,
    "propertySources": [{
        "name": "https://gitee.com/liuge1988/mango.git/src/config-repo/consumer-dev.properties",
        "source": {
            "hello": "hello, dev configurations."
```

```
        }
    }]
}
```

访问 http://localhost:8020/consumer/pro，返回 pro 配置文件的信息。

```
{
    "name": "consumer",
    "profiles": ["pro"],
    "label": null,
    "version": "feab08409500b3626e3f70f3b4d24f7d0cb93efc",
    "state": null,
    "propertySources": [{
        "name": "https://gitee.com/liuge1988/mango.git/src/config-repo/consumer-pro.properties",
        "source": {
            "hello": "hello, pro configurations."
        }
    }]
}
```

上述的返回信息包含了配置文件的位置、版本、配置文件的名称以及配置文件中的具体内容，说明 server 端已经成功获取了 GIT 仓库的配置信息。

访问 http://localhost:8020/consumer-dev.properties，返回结果如图 19-4 所示。

图 19-4

将 dev 配置文件的内容修改为 "hello, dev configurations 2."，如图 19-5 所示。

图 19-5

再次访问 http://localhost:8020/consumer-dev.properties，返回结果如图 19-6 所示。
发现读取的是修改后提交的信息，说明服务端会自动读取最新提交的数据。

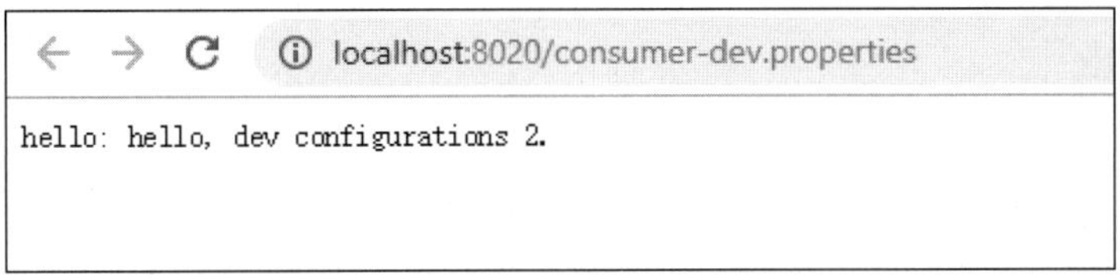

第 19 章　配置中心（Config、Bus）

```
← → C    ⓘ localhost:8020/consumer-dev.properties
hello: hello, dev configurations 2.
```

图 19-6

仓库中的配置文件会被转换成相应的 Web 接口，访问可以参照以下规则：

- /{application}/{profile}[/{label}]
- /{application}-{profile}.yml
- /{label}/{application}-{profile}.yml
- /{application}-{profile}.properties
- /{label}/{application}-{profile}.properties

以 consumer-dev.properties 为例，它的 application 是 consumer、profile 是 dev。客户端会根据填写的参数来选择读取对应的配置。

### 19.3.3　客户端实现

#### 1. 添加依赖

打开 mango-consumer 工程，添加相关依赖。

pom.xml

```xml
<!-- spring-cloud-config -->
<dependency>
    <groupId>org.springframework.cloud</groupId>
    <artifactId>spring-cloud-starter-config</artifactId>
</dependency>
```

#### 2. 配置文件

添加一个 bootstrap.yml 配置文件，添加配置中心，并把注册中心的配置移到这里，因为在通过配置中心查找配置时需要通过注册中心的发现服务。

bootstrap.yml

```yaml
spring:
  cloud:
    consul:
      host: localhost
      port: 8500
      discovery:
        serviceName: ${spring.application.name} # 注册到 consul 的服务名称
    config:
```

```yaml
    discovery:
      enabled: true  # 开启服务发现
      serviceId: mango-config  # 配置中心服务名称
  name: consumer  # 对应{application}部分
  profile: dev  # 对应{profile}部分
  label: master  # 对应git的分支，如果配置中心使用的是本地存储，则该参数无用
```

配置说明：

- spring.cloud.config.uri：配置中心的具体地址。
- spring.cloud.config.name：对应{application}部分。
- spring.cloud.config.profile：对应{profile}部分。
- spring.cloud.config.label：对应git的分支。如果配置中心使用的是本地存储，则该参数无用。
- spring.cloud.config.discovery.service-id：指定配置中心的service-id，便于扩展为高可用配置集群。

注意：上面这些与spring cloud相关的属性必须配置在bootstrap.yml中，这样config部分内容才能被正确加载，因为config的相关配置会先于application.yml，而bootstrap.yml的加载也是先于application.yml文件的。

**application.yml**

```yaml
server:
  port: 8005
spring:
  application:
    name: mango-consumer
  boot:
    admin:
      client:
        url: "http://localhost:8000"
  zipkin:
    base-url: http://localhost:9411/
  sleuth:
    sampler:
      probability: 1   #样本采集量，默认为0.1
# 开放健康检查接口
management:
  endpoints:
    web:
      exposure:
        include: "*"
  endpoint:
```

```yaml
    health:
      show-details: ALWAYS
#开启熔断器
feign:
  hystrix:
    enabled: true
```

### 3. 控制器

添加一个 SpringConfigController 控制器，添加注解 @Value("${hello}")，声明 hello 属性从配置文件读取。

SpringConfigController.java

```java
@RestController
class SpringConfigController {

    @Value("${hello}")
    private String hello;

    @RequestMapping("/hello")
    public String from() {
        return this.hello;
    }

}
```

### 4. 页面测试

启动注册中心、配置中心和服务消费者，访问 http://localhost:8005/hello，返回结果，如图 19-7 所示。

图 19-7

说明客户端已经成功从服务端读取了配置信息。

现在手动修改一下仓库配置文件的内容，移除末尾数字 2，修改完成并提交，如图 19-8 所示。

图 19-8

然后再次访问 http://localhost:80052/hello，返回结果如图 19-9 所示。

图 19-9

我们发现返回结果并没有读取最新提交的内容，这是因为 Spring Boot 项目只有在启动的时候才会获取配置文件的内容，虽然 GIT 配置信息被修改了，但是客户端并没有重新去获取，所以导致读取的信息仍然是旧配置。那么该如何去解决这个问题呢？这就是我们后续要讲的 Spring Cloud Bus。

### 19.3.4　Refresh 机制

我们在上面讲到，Spring Boot 程序只在启动的时候加载配置文件信息，这样在 GIT 仓库配置修改之后，虽然配置中心服务器能够读取最新的提交信息，但是配置中心客户端却不会重新读取，以至于不能及时地读取更新后的配置信息。这时就需要一种通知刷新机制来支持了。

Refresh 机制是 Spring Cloud Config 提供的一种刷新机制，它允许客户端通过 POST 方法触发各自的/refresh，只要依赖 spring-boot-starter-actuator 包就拥有了/refresh 的功能。下面我们为客户端加上刷新功能，以支持更新配置的读取。

**1. 添加依赖**

我们的 mango-consumer 在之前已经添加过 actuator 依赖，所以这里就不用添加了，如果之前没有添加就需要加上。actuator 是健康检查依赖包，依赖包里携带了/refresh 的功能。

```
<dependency>
  <groupId>org.springframework.boot</groupId>
  <artifactId>spring-boot-starter-actuator</artifactId>
</dependency>
```

在使用配置属性的类型中加上 @RefreshScope 注解，这样在客户端执行 /refresh 的时候就会刷新此类下面的配置属性了。

**SpringConfigController.java**

```
@RefreshScope
@RestController
class SpringConfigController {

    @Value("${hello}")
    private String hello;

    @RequestMapping("/hello")
    public String from() {
```

```
        return this.hello;
    }

}
```

**2. 修改配置**

健康检查接口开放需要在配置文件添加以下内容，开放 Refresh 的相关接口，因为这个我们在之前配置过了，所以就不需要添加了。

```
# 开放健康检查接口
management:
  endpoints:
    web:
      exposure:
        include: "*"
  endpoint:
    health:
      show-details: ALWAYS
```

通过上面的接口开放配置，以后以 post 请求的方式访问 http://localhost:8005/actuator/refresh 时就会更新修改后的配置文件了。

注意

这里存在着版本大坑，1.x 跟 2.x 的配置不太一样，我们用的是 2.0+ 版本，务必注意。
（1）安全配置变更
　　新版本为：management.endpoints.web.exposure.include="*"
（2）访问地址变更
　　新版本为：http://localhost:8005/actuator/refresh
　　老版本为：http://localhost:8005/refresh

这里解释一下上面这个配置起到了什么具体作用。其实，actuator 是一个健康检查包，提供了一些健康检查数据接口。Refresh 功能是其中的一个接口，为了安全起见，默认只开放了 health 和 info 接口（启动信息会包含如图 19-10 所示的信息），而上面的配置就是设置要开放哪些接口，我们设置成"*"，是开放所有接口。也可以指定开发几个，比如 health、info、refresh，而这里因为我们需要用的是 Refresh 功能，所以需要把 Refresh 接口开放出来。

```
Exposing 2 endpoint(s) beneath base path '/actuator'
Mapped "{[/actuator/health],methods=[GET],produces=[application/vnd.spri
Mapped "{[/actuator/info],methods=[GET],produces=[application/vnd.spring
Mapped "{[/actuator],methods=[GET],produces=[application/vnd.spring-boot
```

图 19-10

设置成"*"后，启动信息会包含以下信息，这个叫 refresh 的 post 方法就是我们需要的，上面说的接口地址变更从这里也可以看得出来，如图 19-11 所示。

```
Mapped "{[/actuator/archaius],methods=[GET],produces=[application/vnd.spring-boot.actuator.v2+json || appli
Mapped "{[/actuator/auditevents],methods=[GET],produces=[application/vnd.spring-boot.actuator.v2+json || ap
Mapped "{[/actuator/beans],methods=[GET],produces=[application/vnd.spring-boot.actuator.v2+json || applicat
Mapped "{[/actuator/health],methods=[GET],produces=[application/vnd.spring-boot.actuator.v2+json || applica
Mapped "{[/actuator/conditions],methods=[GET],produces=[application/vnd.spring-boot.actuator.v2+json || app
Mapped "{[/actuator/configprops],methods=[GET],produces=[application/vnd.spring-boot.actuator.v2+json || ap
Mapped "{[/actuator/env],methods=[GET],produces=[application/vnd.spring-boot.actuator.v2+json || applicatio
Mapped "{[/actuator/env/{toMatch}],methods=[GET],produces=[application/vnd.spring-boot.actuator.v2+json ||
Mapped "{[/actuator/env],methods=[POST],consumes=[application/vnd.spring-boot.actuator.v2+json || applicati
Mapped "{[/actuator/env],methods=[DELETE],produces=[application/vnd.spring-boot.actuator.v2+json || applica
Mapped "{[/actuator/info],methods=[GET],produces=[application/vnd.spring-boot.actuator.v2+json || applicati
Mapped "{[/actuator/loggers],methods=[GET],produces=[application/vnd.spring-boot.actuator.v2+json || applic
Mapped "{[/actuator/loggers/{name}],methods=[GET],produces=[application/vnd.spring-boot.actuator.v2+json ||
Mapped "{[/actuator/loggers/{name}],methods=[POST],consumes=[application/vnd.spring-boot.actuator.v2+json |
Mapped "{[/actuator/heapdump],methods=[GET],produces=[application/octet-stream]}" onto public java.lang.Obj
Mapped "{[/actuator/threaddump],methods=[GET],produces=[application/vnd.spring-boot.actuator.v2+json || app
Mapped "{[/actuator/metrics],methods=[GET],produces=[application/vnd.spring-boot.actuator.v2+json || applic
Mapped "{[/actuator/metrics/{requiredMetricName}],methods=[GET],produces=[application/vnd.spring-boot.actua
Mapped "{[/actuator/scheduledtasks],methods=[GET],produces=[application/vnd.spring-boot.actuator.v2+json ||
Mapped "{[/actuator/httptrace],methods=[GET],produces=[application/vnd.spring-boot.actuator.v2+json || appl
Mapped "{[/actuator/mappings],methods=[GET],produces=[application/vnd.spring-boot.actuator.v2+json || appli
Mapped "{[/actuator/refresh],methods=[POST],produces=[application/vnd.spring-boot.actuator.v2+json || appli
Mapped "{[/actuator/features],methods=[GET],produces=[application/vnd.spring-boot.actuator.v2+json || appli
Mapped "{[/actuator/service-registry],methods=[GET],produces=[application/vnd.spring-boot.actuator.v2+json
Mapped "{[/actuator/service-registry],methods=[POST],consumes=[application/vnd.spring-boot.actuator.v2+jsor
Mapped "{[/actuator/consul],methods=[GET],produces=[application/vnd.spring-boot.actuator.v2+json || applica
Mapped "{[/actuator],methods=[GET],produces=[application/vnd.spring-boot.actuator.v2+json || application/js
```

图 19-11

### 3. 页面测试

重新启动服务，访问 http://localhost:8005/hello，返回结果如图 19-12 所示。

图 19-12

修改仓库配置内容，末尾加个数字 5，如图 19-13 所示。

图 19-13

再次访问 http://localhost:8005/hello，如我们所料，结果并没有更新，因为我们还没有调 refresh 方法。通过工具或自写代码发送 post 请求 http://localhost:8005/actuator/refresh，刷新配置。这里通过在线测试网站发送，地址为 https://getman.cn/Mo2FX 。

注意 先让你的 Chrome 支持跨域。设置方法是，在快捷方式的 target 后加上 --disable-web-security --user-data-dir，重启即可，如图 19-14 所示。

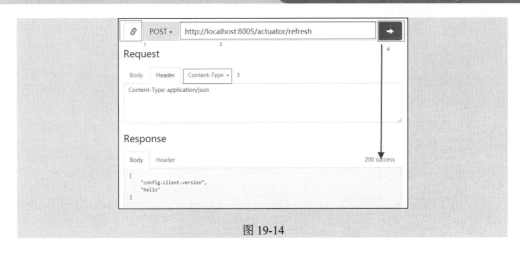

图 19-14

刷新之后，再次访问 http://localhost:8005/hello，返回结果如图 19-15 所示。

图 19-15

查看返回结果，刷新之后已经可以获取最新提交的配置内容，但是每次都需要手动刷新客户端还是很麻烦，如果客户端数量一多就简直难以忍受了，有没有什么比较好的办法来解决这个问题呢？那是当然的，答案就是：Spring Cloud Bus。

## 19.3.5　Spring Cloud Bus

Spring Cloud Bus 被大家称为消息总线，通过轻量级的消息代理来连接各个分布的节点，可以利用像消息队列的广播机制在分布式系统中进行消息传播。通过消息总线可以实现很多业务功能，其中对于配置中心客户端刷新就是一个非常典型的使用场景。图 19-16 可以很好地解释消息总线的作用流程。

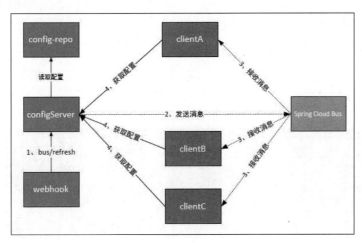

图 19-16

Spring Cloud Bus 进行配置更新的步骤如下：

（1）提交代码触发 post 请求给/actuator/bus-refresh。
（2）Server 端接收到请求并发送给 Spring Cloud Bus。
（3）Spring Cloud bus 接到消息并通知给其他客户端。
（4）其他客户端接收到通知，请求 Server 端获取最新配置。
（5）全部客户端均获取到最新的配置。

### 1. 安装 RabbitMQ

因为我们需要用到消息队列，所以这里选择 RabbitMQ，使用 Docker 进行安装。

（1）拉取镜像

执行以下命令，拉取镜像：

```
docker pull rabbitmq:management
```

完成之后执行以下命令查看下载镜像：

```
docker images
```

查看镜像列表，如图 19-17 所示。

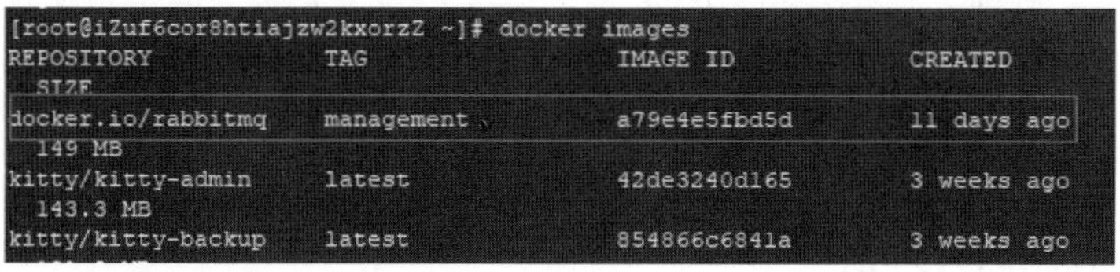

图 19-17

（2）创建容器

执行以下命令，创建 Docker 容器：

```
docker run -d --name rabbitmq -p 5671:5671 -p 5672:5672 -p 4369:4369 -p 25672:25672 -p 15671:15671 -p 15672:15672 rabbitmq:management
```

启动成功之后，可以执行以下命令查看启动容器：

```
docker ps
```

查看启动容器，如图 19-18 所示。

图 19-18

## （3）登录界面

容器启动之后就可以访问 Web 管理界面了（访问 http://宿主机 IP:15672）。
系统提供了默认账号：用户名为 guest，密码为 guest，如图 19-19 所示。

图 19-19

管理界面如图 19-20 所示。

### 2．客户端实现

（1）添加依赖

打开客户端 mango-consumer，添加消息总线相关依赖。

图 19-20

**pom.xml**

```xml
<!-- bus-amqp -->
<dependency>
    <groupId>org.springframework.cloud</groupId>
    <artifactId>spring-cloud-starter-bus-amqp</artifactId>
</dependency>
```

（2）修改配置

添加 RebbitMQ 的相关配置，这样客户端代码就改造完成了。

bootstrap.yml

```yaml
spring:
  cloud:
    consul:
      host: localhost
      port: 8500
      discovery:
        serviceName: ${spring.application.name} # 注册到consul的服务名称
    config:
      discovery:
        enabled: true  # 开启服务发现
        serviceId: mango-config # 配置中心服务名称
      name: consumer  # 对应{application}部分
      profile: dev  # 对应{profile}部分
      label: master  # 对应git的分支，如果配置中心使用的是本地存储，则该参数无用
  rabbitmq:
    host: localhost
    port: 5672
    username: guest
    password: guest
```

### 3. 服务端实现

（1）添加依赖

修改 mango-config，添加相关依赖。

pom.xml

```xml
<dependency>
    <groupId>org.springframework.boot</groupId>
    <artifactId>spring-boot-starter-actuator</artifactId>
</dependency>
<dependency>
    <groupId>org.springframework.cloud</groupId>
    <artifactId>spring-cloud-starter-bus-amqp</artifactId>
</dependency>
```

（2）修改配置

添加 RabbitMQ 和接口开放相关的配置，这样服务端代码就改造完成了。

application.yml

```yaml
spring:
  rabbitmq:
    host: localhost
```

```
      port: 5672
      username: guest
      password: guest
```

**4．页面测试**

（1）启动服务端，成功集成消息总线后，启动信息中可以看到如图 19-21 所示的信息。

```
Mapped "{[/actuator/refresh],methods=[POST],produces=[application/vnd.spring-boot.actuato
Mapped "{[/actuator/bus-env/{destination}],methods=[POST],consumes=[application/vnd.sprin
Mapped "{[/actuator/bus-env],methods=[POST],consumes=[application/vnd.spring-boot.actuato
Mapped "{[/actuator/bus-refresh/{destination}],methods=[POST]}" onto public java.lang.Obj
Mapped "{[/actuator/bus-refresh],methods=[POST]}" onto public java.lang.Object org.spring
Mapped "{[/actuator/features],methods=[GET],produces=[application/vnd.spring-boot.actuato
Mapped "{[/actuator/service-registry],methods=[POST],consumes=[application/vnd.spring-boo
Mapped "{[/actuator/service-registry],methods=[GET],produces=[application/vnd.spring-boot
```

图 19-21

（2）启动客户端，发现居然报错了。在网上找不到相关资料，也没见其他人提过相关问题，猜测网上教程多是使用 Euraka，而这里用的是 Consul。不想换回 Euraka，2.0 停止开发消息出来以后，将来还不一定是什么情况，就只能硬着头皮解决了。

```
org.springframework.beans.factory.NoSuchBeanDefinitionException: No bean named
'configServerRetryInterceptor' available
    at org.springframework.beans.factory.support.DefaultListableBeanFactory.
getBeanDefinition(DefaultListableBeanFactory.java:685)
~[spring-beans-5.0.8.RELEASE.jar:5.0.8.RELEASE]
    at org.springframework.beans.factory.support.AbstractBeanFactory.
getMergedLocalBeanDefinition(AbstractBeanFactory.java:1210)
~[spring-beans-5.0.8.RELEASE.jar:5.0.8.RELEASE]
    at org.springframework.beans.factory.support.AbstractBeanFactory.doGetBean
(AbstractBeanFactory.java:291) ~[spring-beans-5.0.8.RELEASE.jar:5.0.8.RELEASE]
    at org.springframework.beans.factory.support.AbstractBeanFactory.getBean
(AbstractBeanFactory.java:204) ~[spring-beans-5.0.8.RELEASE.jar:5.0.8.RELEASE]
    at org.springframework.retry.annotation.
AnnotationAwareRetryOperationsInterceptor.getDelegate(AnnotationAwareRetryOper
ationsInterceptor.java:180) ~[spring-retry-1.2.2.RELEASE.jar:na]
    at org.springframework.retry.annotation.
AnnotationAwareRetryOperationsInterceptor.invoke(AnnotationAwareRetryOperation
sInterceptor.java:151) ~[spring-retry-1.2.2.RELEASE.jar:na]
```

然后跟踪代码，发现在图 19-22 中的位置找不到相应的 Bean 了，答案也就比较明显了：要么是程序有 BUG（不过可能性不大），要么就是缺包了，在缺失的包里有这个 Bean。但是这个 Bean 在哪个包里呢？网上没有详细资料，在网上对比了一下消息总线的配置，依赖也没有少加什么，如图 19-22 所示。

```
public class ConfigServerInstanceProvider {

    private static Log logger = LogFactory.getLog(ConfigServerInstanceProvider.class);
    private final DiscoveryClient client;

    public ConfigServerInstanceProvider(DiscoveryClient client) {
        this.client = client;
    }
                                            找不到对应的Bean
    @Retryable(interceptor = "configServerRetryInterceptor")
    public List<ServiceInstance> getConfigServerInstances(String serviceId) {
        logger.debug("Locating configserver (" + serviceId + ") via discovery");
        List<ServiceInstance> instances = this.client.getInstances(serviceId);
        if (instances.isEmpty()) {
            throw new IllegalStateException(
                    "No instances found of configserver (" + serviceId + ")");
        }
        logger.debug("Located configserver (" + serviceId
                + ") via discovery. No of instances found: " + instances.size());
        return instances;
    }
}
```

图 19-22

（3）在刷新的时候缺少一个拦截器，可以自己设置一个。加一个配置类，并在 resources 下新建 META-INF 目录和一个 spring.factories 文件，如图 19-23 所示。

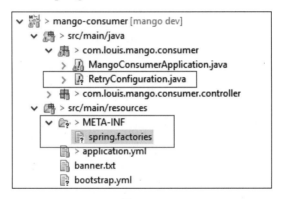

图 19-23

### RetryConfiguration.java

```
public class RetryConfiguration {

    @Bean
    @ConditionalOnMissingBean(name = "configServerRetryInterceptor")
    public RetryOperationsInterceptor configServerRetryInterceptor() {
        return RetryInterceptorBuilder.stateless().backOffOptions(1000, 1.2, 5000).maxAttempts(10).build();
    }
}
```

### spring.factories

```
org.springframework.cloud.bootstrap.BootstrapConfiguration=com.louis.mango.consumer.RetryConfiguration
```

指定新建的拦截器,这样系统初始化时就会加载这个 Bean。然后重新启动,就没有报错了。下面看看能不能使用。

(4) 访问 http://localhost:8005/hello,效果如图 19-24 所示。

图 19-24

(5) 修改仓库配置文件,把数字 5 改成 15,修改完成提交,如图 19-25 所示。

图 19-25

再次访问发现还是旧信息,如图 19-26 所示。

图 19-26

(6) 用工具发送 post 请求 http://localhost:8020/actuator/bus-refresh。

 这次是向注册中心服务端发送请求,发送成功之后服务端会通过消息总线通知所有的客户端进行刷新。另外,开启消息总线后的请求地址是 /actuator/bus-refresh,不再是 refresh 了,如图 19-27 所示。

图 19-27

（7）给服务端发送刷新请求之后，再次访问 http://localhost:8005/hello，结果如图 19-28 所示（可能需要一点刷新时间）。

图 19-28

最终，我们愉快地发现客户端已经能够通过消息总线获取最新配置了。

# 第三篇
# 前端实现篇

本篇内容为前端实现篇，全面细致地讲解 Mango 权限管理系统的前端实现全过程。从零开始，逐步扩展，逐渐完善，手把手地教你如何利用 Vue.js 和 Element 构建功能丰富、风格优雅的权限管理系统。

第 20 章 搭建开发环境，完整地阐述和示范前端开发环境的搭建和安装。
第 21 章 前端项目案例，讲解基于 Vue + Element 实现的第一个案例。
第 22 章 工具模块封装，对常用的 axios 和 Mock 模块进行集中封装。
第 23 章 第三方图标库，介绍第三方图标库 Font Awesome 的使用方法。
第 24 章 多语言国际化，讲解如何通过 Vue 组件实现多语言国际化。
第 25 章 登录流程完善，丰富登录功能，美化登录界面，优化登录逻辑。
第 26 章 管理应用状态，讲解如何使用 vuex 进行组件间状态的共享。
第 27 章 头部功能组件，实现和优化主题切换、语言切换等功能组件。
第 28 章 动态加载菜单，讲解如何通过导航守卫动态加载导航菜单树。
第 29 章 页面权限控制，讲解页面和按钮权限控制的实现思路和方案。
第 30 章 功能管理模块，抽取部分案例讲解常规业务功能模块页面的实现。
第 31 章 嵌套外部网页，讲解使用 IFrame 嵌套外部网页的实现思路和方案。
第 32 章 数据备份还原，讲解数据备份还原前端界面相关的实现方案。

# 第 20 章 搭建开发环境

## 20.1 技术基础

前端项目将会使用到以下几项主要技术和框架，若有必要，请先了解相关知识，也可一边跟进项目一边补充学习，请读者自行斟酌。

- Vue.js 官网：https://cn.vuejs.org/
- Vue.js 教程：http://www.runoob.com/vue2/vue-tutorial.html
- Vue-router 教程：https://router.vuejs.org/zh/
- Vuex 教程：https://vuex.vuejs.org/zh/guide/
- Element 教程：http://element-cn.eleme.io/#/zh-CN

## 20.2 开发环境

### 20.2.1 Visual Studio Code

Visual Studio Code 是一款非常优秀的开源编辑器，非常适合作为前端 IDE，根据自己的系统下载相应的版本进行安装，如图 20-1 所示。

- 下载地址：https://code.visualstudio.com/

更多 VS Code 教程可以参考以下资料：

- 官网文档：https://code.visualstudio.com/docs
- 简书教程：https://www.jianshu.com/p/990b19834896

图 20-1

### 20.2.2　Node JS

Node JS 提供的 NPM 依赖管理和编译打包工具使用起来非常方便，对于前端比较大型一些的项目还是采用 NPM 作为打包工具比较理想。要使用 NPM，就需要安装 NodeJS，下载地址为 http://nodejs.cn/download/。

选择系统对应的版本，这里我们下载 Windows 系统的 64 位 zip 文件 node-v10.15.0-win-x64.zip。下载完成后解压，可以看到里面有一个 node.exe 的可执行文件，如图 20-2 所示。

把 Node 添加到系统环境变量里面，打开 cmd 命令行，输入 npm -v，如果出现如图 20-3 所示的提示信息，就说明已经安装正确。

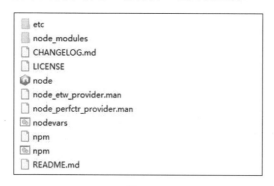

图 20-2　　　　　　　　　　　　　　图 20-3

如果你安装的是旧版本的 npm，可以很容易地通过 npm 命令来升级。

```
# linux 系统命令
sudo npm install npm -g
# windows 系统命令
npm install npm -g
```

如图 20-4 所示，npm 从 6.4.1 版本升级到了 6.5.0 版本。

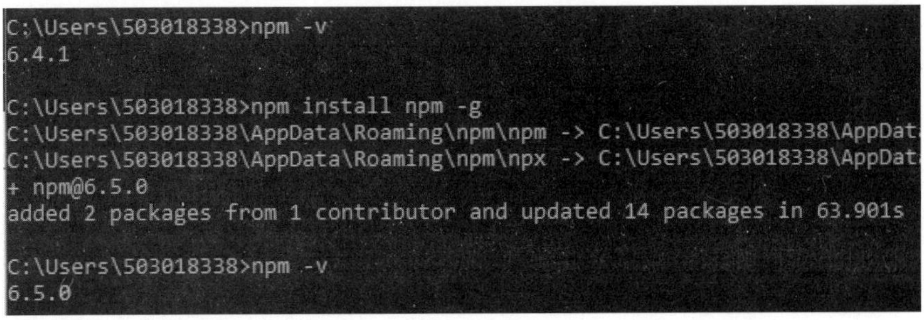

```
C:\Users\503018338>npm -v
6.4.1

C:\Users\503018338>npm install npm -g
C:\Users\503018338\AppData\Roaming\npm\npm -> C:\Users\503018338\AppDat
C:\Users\503018338\AppData\Roaming\npm\npx -> C:\Users\503018338\AppDat
+ npm@6.5.0
added 2 packages from 1 contributor and updated 14 packages in 63.901s

C:\Users\503018338>npm -v
6.5.0
```

图 20-4

更多 NodeJS 教程可以参考以下资料：
- 中文官网：http://nodejs.cn/api/
- 菜鸟学堂：https://www.runoob.com/nodejs/nodejs-tutorial.html

### 20.2.3 安装 webpack

安装好 npm 之后，就可以通过 npm 命令来下载各种工具了。安装打包工具 webpack，-g 表示全局安装，代码如下：

```
npm install webpack -g
```

更多 webpack 教程可以参考以下资料：
- 菜鸟学堂：http://www.runoob.com/w3cnote/webpack-tutorial.html

### 20.2.4 安装 vue-cli

安装 vue 脚手架项目初始化工具 vue-cli，-g 表示全局安装，代码如下：

```
npm install vue-cli -g
```

### 20.2.5 淘宝镜像

因为 NPM 使用的是国外中央仓库，有时候下载速度比较"喜人"，就像 Maven 有国内镜像一样，NPM 在国内也有镜像可用。这里建议使用淘宝镜像。

安装淘宝镜像，安装成功后用 cnpm 替代 npm 命令即可，如 cnpm install webpack -g 。

```
npm install -g cnpm --registry=https://registry.npm.taobao.org
```

### 20.2.6 安装 Yarn

Yarn 是 Facebook 发布的 node.js 包管理器，比 npm 更快、更高效，可以使用 Yarn 替代 npm。安装了 Node，同时也就安装了 NPM，可以使用下面的命令来安装：

```
npm i yarn -g -verbose
```

NPM 官方源访问速度实在不敢恭维，建议使用之前切换为淘宝镜像，在 Yarn 安装完毕之后执行如下指令：

```
yarn config set registry https://registry.npm.taobao.org
```

到此为止，我们就可以在项目中像使用 NPM 一样使用 Yarn 了。使用 Yarn 跟 NPM 差别不大，具体命令关系如下：

```
npm install  => yarn install
npm install --save [package] => yarn add [package]
npm install --save-dev [package] => yarn add [package] --dev
npm install --global [package] => yarn global add [package]
npm uninstall --save [package] => yarn remove [package]
npm uninstall --save-dev [package] => yarn remove [package]
```

# 20.3 创建项目

## 20.3.1 生成项目

环境已经搭建完成，现在我们通过 vue-cli 来生成一个项目，名称为 mango-ui。

```
vue init webpack mango-ui
```

一路根据提示输入项目信息，等待项目生成，如图 20-5 所示。

命令执行完毕，生成的项目结构如图 20-6 所示。

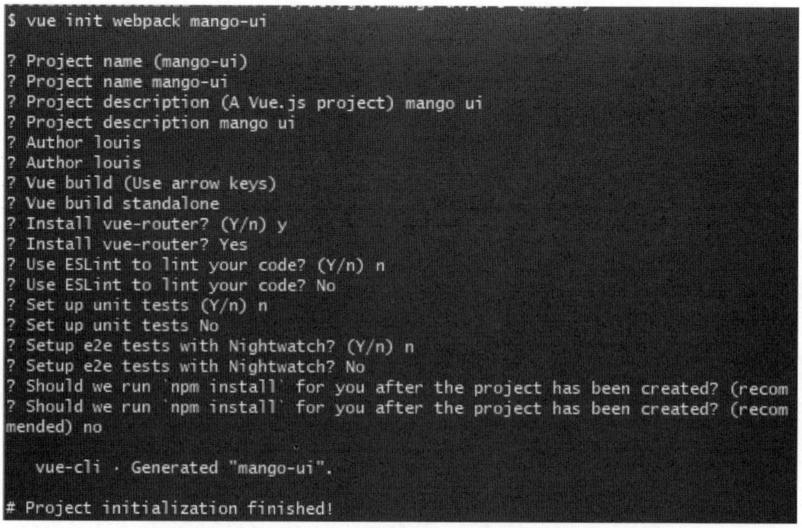

```
$ vue init webpack mango-ui

? Project name (mango-ui)
? Project name mango-ui
? Project description (A Vue.js project) mango ui
? Project description mango ui
? Author louis
? Author louis
? Vue build (Use arrow keys)
? Vue build standalone
? Install vue-router? (Y/n) y
? Install vue-router? Yes
? Use ESLint to lint your code? (Y/n) n
? Use ESLint to lint your code? No
? Set up unit tests (Y/n) n
? Set up unit tests No
? Setup e2e tests with Nightwatch? (Y/n) n
? Setup e2e tests with Nightwatch? No
? Should we run `npm install` for you after the project has been created? (recom
? Should we run `npm install` for you after the project has been created? (recom
mended) no

   vue-cli · Generated "mango-ui".

# Project initialization finished!
```

图 20-5

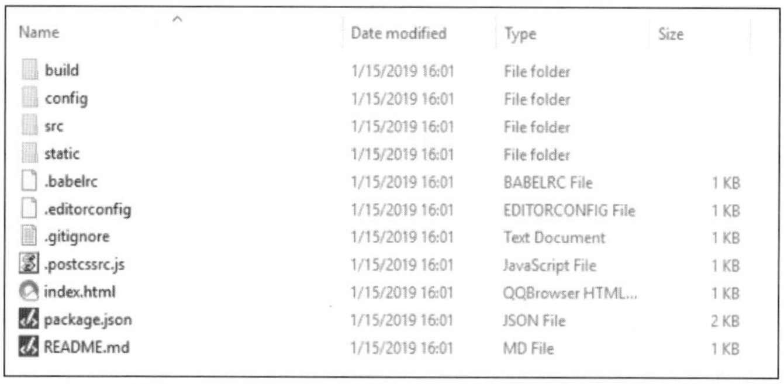

图 20-6

## 20.3.2 安装依赖

进入项目根目录，执行 yarn install（也可以用 npm install 或淘宝 cnpm install，我们这里用 Yarn 会快一点）安装依赖包。

```
cd mango-ui
yarn install
```

安装成功之后如图 20-7 所示。

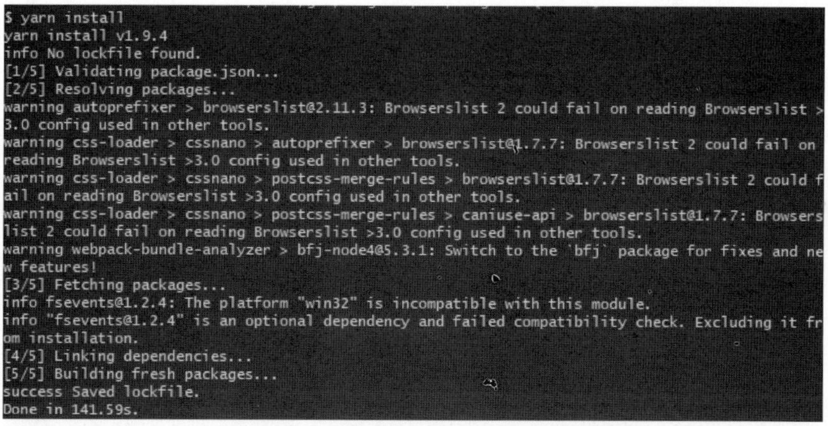

图 20-7

等依赖包安装完成之后，会在项目根目录下生成 node_modules 文件夹，这个目录就是下载的依赖包的统一存放目录，如图 20-8 所示。

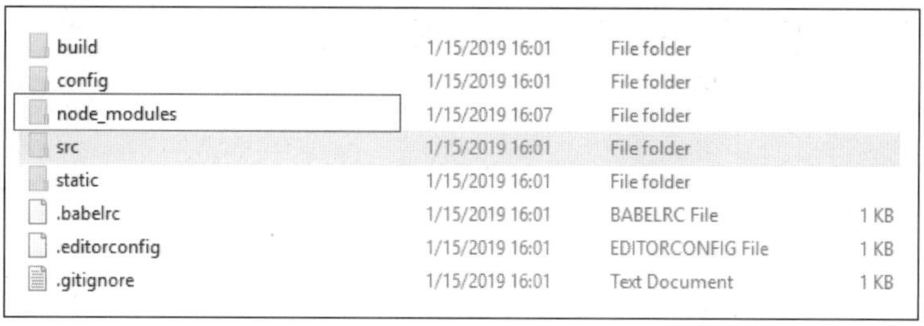

图 20-8

### 20.3.3 启动运行

安装完成之后，执行应用启动命令，运行项目：

```
npm run dev
```

执行命令之后，如果显示 "I Your application is runing here ...."，就表示启动成功了，如图 20-9 所示。

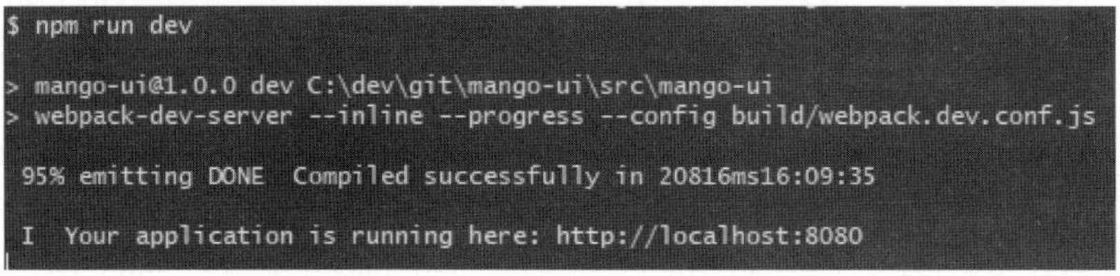

图 20-9

浏览器访问对应地址，如图 20-10 所示，这里是 http://localhost:8080，会出现 Vue 的介绍页面。

图 20-10

到此，我们的项目脚手架就建立起来了。

# 第 21 章 前端项目案例

## 21.1 导入项目

打开 Visual Studio Code，选择 File→add Folder to Workspace，导入我们的项目，如图 21-1 所示。

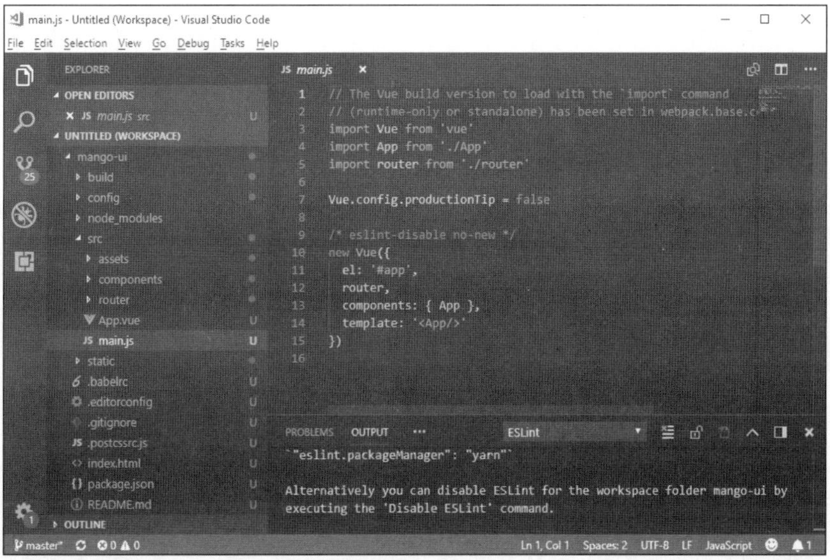

图 21-1

## 21.2 安装 Element

### 21.2.1 安装依赖

Element 是国内饿了么公司提供的一套开源前端框架，简洁优雅，提供了 Vue、React、Angular 等多个版本，这里使用 Vue 版本来搭建我们的界面。

访问地址 http://element-cn.eleme.io/#/zh-CN/component/installation，查看官方指南，包含框架的安装、组件的使用等全方位的教程，如图 21-2 所示。

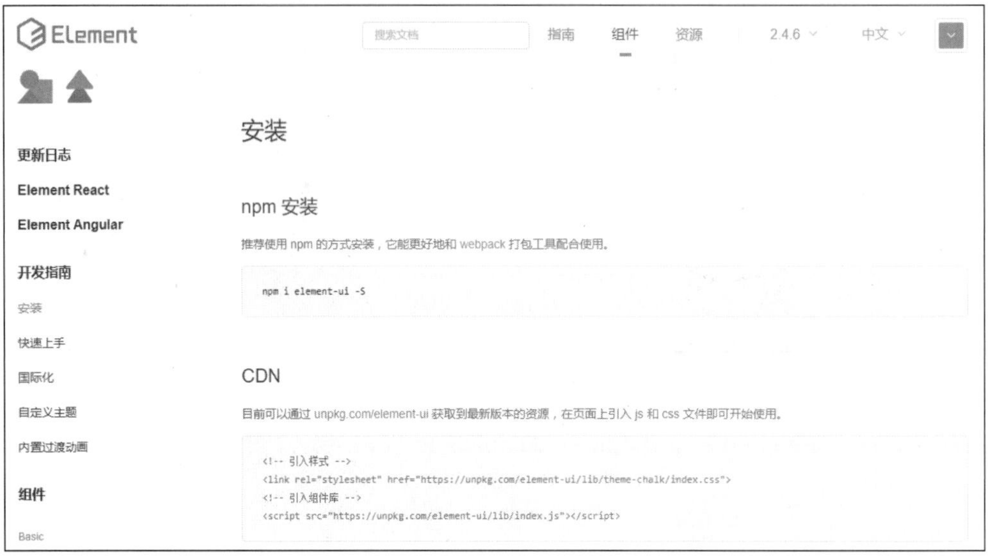


图 21-2

按照安装指南，我们选择 NPM 的安装方式。因为我们安装了 Yarn，所以可以直接使用 yarn add element-ui 命令替代，如图 21-3 所示。

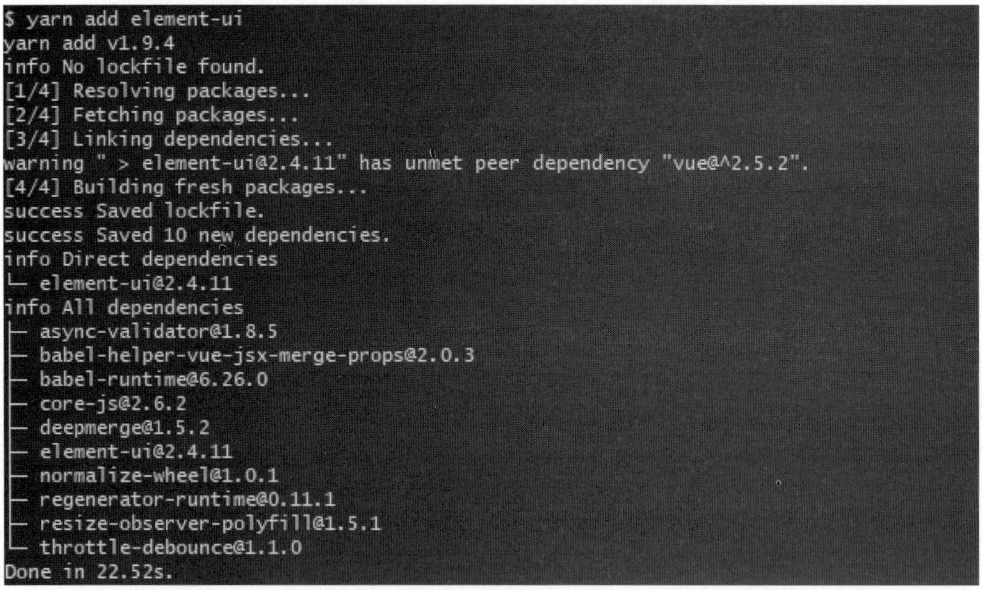
```
$ yarn add element-ui
yarn add v1.9.4
info No lockfile found.
[1/4] Resolving packages...
[2/4] Fetching packages...
[3/4] Linking dependencies...
warning " > element-ui@2.4.11" has unmet peer dependency "vue@^2.5.2".
[4/4] Building fresh packages...
success Saved lockfile.
success Saved 10 new dependencies.
info Direct dependencies
└─ element-ui@2.4.11
info All dependencies
├─ async-validator@1.8.5
├─ babel-helper-vue-jsx-merge-props@2.0.3
├─ babel-runtime@6.26.0
├─ core-js@2.6.2
├─ deepmerge@1.5.2
├─ element-ui@2.4.11
├─ normalize-wheel@1.0.1
├─ regenerator-runtime@0.11.1
├─ resize-observer-polyfill@1.5.1
└─ throttle-debounce@1.1.0
Done in 22.52s.
```

图 21-3

## 21.2.2 导入项目

按照安装指南，在 main.js 中引入 Element。引入之后，main.js 内容如图 21-4 所示。

```
import Vue from 'vue'
import App from './App'
import router from './router'
import ElementUI from 'element-ui'
import 'element-ui/lib/theme-chalk/index.css'

Vue.config.productionTip = false

Vue.use(ElementUI)

new Vue({
  el: '#app',
  router,
  render: h => h(App)
})
```

图 21-4

引入项目之后，在原有的 HelloWorld.vue 页面中加入一个 Element 的按钮，测试一下。在 Element 官网组件教程案例中，包含大量组件使用场景，直接复制组件代码到项目页面即可，如图 21-5 所示。

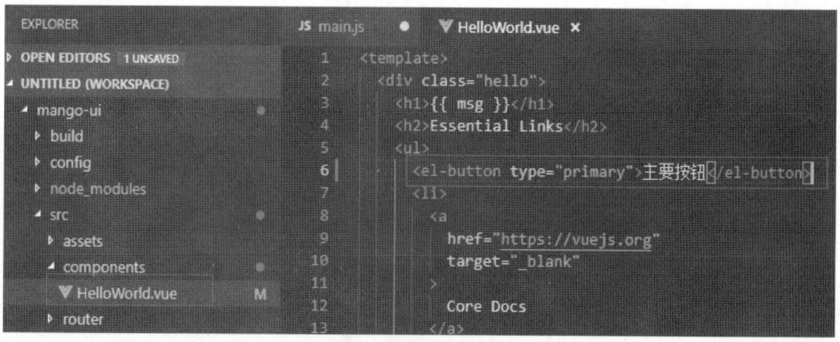


图 21-5

如图 21-6 所示，说明 Element 组件已经成功引入了。

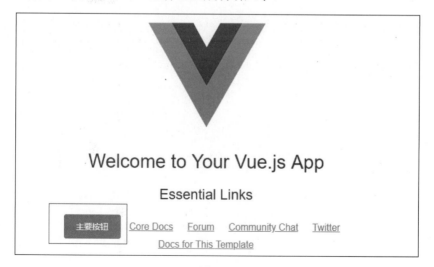

图 21-6

## 21.3 页面路由

### 21.3.1 添加页面

我们把 components 改名为 views,并在 views 目录下添加 3 个页面:Login.vue、Home.vue、404.vue。3 个页面内容简单相似,只有简单的页面标识,如登录页面是"Login Page"。Login.vue 代码如下,其他页面类似。

```
<template>
  <div class="page">
    <h2>Login Page</h2>
  </div>
</template>

<script>
export default {
  name: 'Login'
}
</script>
```

### 21.3.2 配置路由

打开 router/index.js,添加 3 个路由,分别对应主页、登录和 404 页面。

```
import Vue from 'vue'
import Router from 'vue-router'
import Login from '@/views/Login'
import Home from '@/views/Home'
import NotFound from '@/views/404'

Vue.use(Router)

export default new Router({
  routes: [
    {
      path: '/',
      name: 'Home',
      component: Home
    }, {
      path: '/login',
```

```
    name: 'Login',
    component: Login
  }, {
    path: '/404',
    name: 'notFound',
    component: NotFound
  }
  ]
})
```

在浏览器中重新访问下面几个不同的路径，路由器会根据路径路由到相应的页面。

- http://localhost:8080/#/，/ 路由到 Home Page，如图 21-7 所示。

图 21-7

- http://localhost:8080/#/login，/login 路由到 Login Page，如图 21-8 所示。

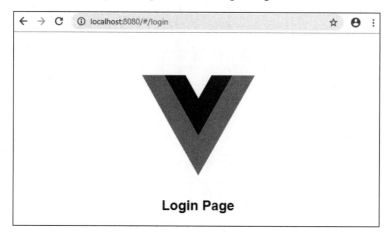

图 21-8

- http://localhost:8080/#/404，/404 路由到 404 Page，如图 21-9 所示。

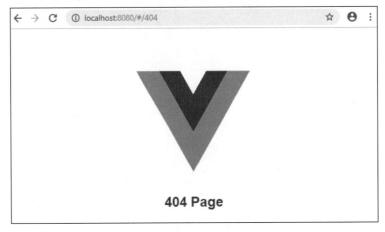

图 21-9

## 21.4 安装 SCSS

### 21.4.1 安装依赖

因为后续会用到 SCSS 编写页面样式，所以先安装好 SCSS。

```
yarn add sass-loader node-sass -dev
```

### 21.4.2 添加配置

在 build 文件夹下的 webpack.base.conf.js 的 rules 标签下添加配置。

```
{
    test: /\.scss$/,
    loaders: ['style', 'css', 'sass']
}
```

### 21.4.3 如何使用

在页面代码 style 标签中把 lang 设置成 scss 即可。

```
<style lang="scss">
</style>
```

### 21.4.4 页面测试

丰富一下 404 页面内容，加入 scss 样式，如图 21-10 所示。页面样式代码太多，不方便贴，详见代码。

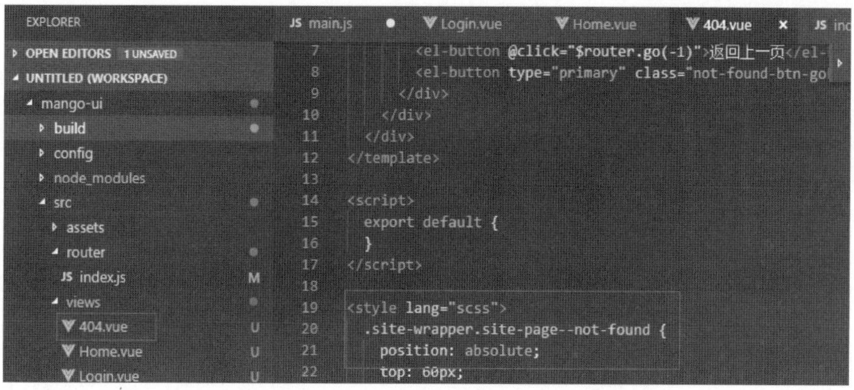

图 21-10

访问 http://localhost:8080/#/404，正确显示修改后的 404 页面效果，如图 21-11 所示。

图 21-11

# 21.5 安装 axios

axios 是一个基于 Promise 用于浏览器和 Node.js 的 HTTP 客户端，我们后续需要用来发送 HTTP 请求，接下讲解 axios 的安装和使用。

## 21.5.1 安装依赖

执行以下命令，安装 axios 依赖。

```
yarn add axios
```

## 21.5.2 编写代码

安装完成后，修改 Home.vue，进行简单的安装测试。

```
<template>
  <div class="page">
    <h2>Home Page</h2>
    <el-button type="primary" @click="testAxios()">测试 Axios 调用</el-button>
  </div>
</template>

<script>
import axios from 'axios'
export default {
  name: 'Home',
  methods: {
    testAxios() {
      axios.get('http://localhost:8080').then(res => { alert(res.data) })
    }
  }
}
</script>
```

### 21.5.3 页面测试

打开主页，单击测试按钮触发 HTTP 请求，并弹出窗显示返回页面的 HTML 数据，如图 21-12 所示。

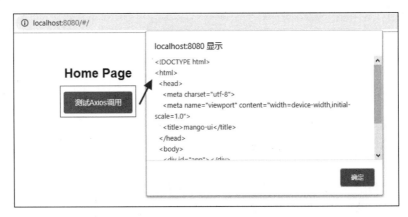

图 21-12

# 21.6 安装 Mock.js

为了模拟后台接口提供页面需要的数据，引入 Mock.js 为我们提供模拟数据，而不用依赖于后台接口的完成。

## 21.6.1 安装依赖

执行如下命令，安装依赖包。

```
yarn add mockjs -dev
```

## 21.6.2 编写代码

安装完成之后，我们写个例子测试一下。

在 src 目录下新建一个 mock 目录，创建 mock.js，在里面模拟两个接口，分别拦截用户和菜单的请求，并返回相应的数据。

```
import Mock from 'mockjs'
Mock.mock('http://localhost:8080/user', {
    'name': '@name', // 随机生成姓名
    'name': '@email', // 随机生成邮箱
    'age|1-10': 5, // 年龄在 1~10 岁之间
})
Mock.mock('http://localhost:8080/menu', {
    'id': '@increment', // id自增
    'name': 'menu', // 名称为 menu
    'order|1-20': 5, // 排序在 1~20 之间
})
```

修改 Home.vue，在页面添加两个按钮，分别触发用户和菜单的处理请求，成功后弹出获取结果。

注意需要在页面通过 import mock from '@/mock/mock.js' 语句引入 mock 模块。

```
<template>
  <div class="page">
    <h2>Home Page</h2>
    <el-button type="primary" @click="testAxios()">测试 Axios 调用</el-button>
    <el-button type="primary" @click="getUser()">获取用户信息</el-button>
    <el-button type="primary" @click="getMenu()">获取菜单信息</el-button>
  </div>
</template>

<script>
import axios from 'axios'
import mock from '@/mock/mock.js'
export default {
  name: 'Home',
  methods: {
    testAxios() {
```

```
      axios.get('http://localhost:8080').then(res => { alert(res.data) })
    },
    getUser() {
      axios.get('http://localhost:8080/user').then(res => 
{ alert(JSON.stringify(res.data)) })
    },
    getMenu() {
      axios.get('http://localhost:8080/menu').then(res => 
{ alert(JSON.stringify(res.data)) })
    }
  }
}
</script>
```

### 21.6.3 页面测试

在浏览器中访问 http://localhost:8080/#/，分别单击两个按钮，mock 会根据请求 url 拦截对应请求并返回模拟数据。

- 获取用户信息，如图 21-13 所示。

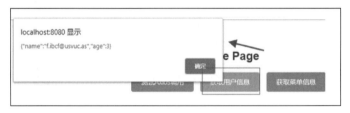

图 21-13

- 获取菜单信息，如图 21-14 所示。

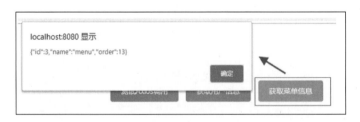

图 21-14

测试成功，这样 mock 就成功集成进来了。

# 第 22 章 工具模块封装

## 22.1 封装 axios 模块

### 22.1.1 封装背景

使用 axios 发起一个请求是比较简单的事情，但是 axios 没有进行封装复用，项目越来越大，会引起越来越多的代码冗余，让代码变得越来越难维护，所以我们在这里先对 axios 进行二次封装，使项目中各个组件能够复用请求，让代码变得更容易维护。

### 22.1.2 封装要点

- 统一 url 配置。
- 统一 api 请求。
- request（请求）拦截器。例如：带上 token 等，设置请求头。
- response（响应）拦截器。例如：统一错误处理，页面重定向等。
- 根据需要，结合 vuex 做全局的 loading 动画，或者错误处理。
- 将 axios 封装成 Vue 插件使用。

### 22.1.3 文件结构

在 src 目录下，新建一个 http 文件夹，用来存放 http 交互 api 代码。文件结构如下：

- config.js：axios 默认配置，包含基础路径等信息。
- axios.js：二次封装 axios 模块，包含拦截器等信息。
- api.js：请求接口汇总模块，聚合所有模块 API。
- index.js：将 axios 封装成插件，按插件方式引入。
- modules：用户管理、菜单管理等子模块 API。

http 模块目录文件结构图如图 22-1 所示。
子模块 Modules 目录文件结构图如图 22-2 所示。

图 22-1

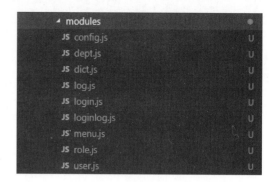

图 22-2

## 22.1.4 代码说明

### 1. config.js

AXIOS 相关配置，每个配置项都有带说明，详见注释。

```
import { baseUrl } from '@/utils/global'

export default {
  method: 'get',
  // 基础 url 前缀
  baseUrl: baseUrl,
  // 请求头信息
  headers: {
    'Content-Type': 'application/json;charset=UTF-8'
  },
  // 参数
  data: {},
  // 设置超时时间
  timeout: 10000,
  // 携带凭证
  withCredentials: true,
  // 返回数据类型
  responseType: 'json'
}
```

### 2. axios.js

axios 拦截器，可以进行请求拦截和响应拦截，在发送请求和响应请求时执行一些操作。

（1）这里导入类配置文件的信息（如 baseURL、headers、 withCredentials 等设置）到 axios 对象。

（2）发送请求的时候获取 token，如果 token 不存在，说明未登录，就重定向到系统登录界面，否则携带 token 继续发送请求。

（3）如果有需要，可以在这里通过 response 响应拦截器对返回结果进行统一处理后再返回。

```js
import axios from 'axios';
import config from './config';
import Cookies from "js-cookie";
import router from '@/router'

export default function $axios(options) {
  return new Promise((resolve, reject) => {
    const instance = axios.create({
      baseURL: config.baseUrl,
      headers: config.headers,
      timeout: config.timeout,
      withCredentials: config.withCredentials
    })
    // request 请求拦截器
    instance.interceptors.request.use(
      config => {
        let token = Cookies.get('token')
        if (token) {   // 发送请求时携带 token
          config.headers.token = token
        } else {   // 重定向到登录页面
          router.push('/login')
        }
        return config
      },
      error => {
        return Promise.reject(error)
      }
    )

    // response 响应拦截器
    instance.interceptors.response.use(
      response => {
        return response.data
      },
      err => {
        return Promise.reject(err)
      }
    )
```

```
    // 请求处理
    instance(options).then(res => {
      resolve(res)
      return false
    }).catch(error => {
      reject(error)
    })
  })
}
```

### 3. index.js

这里把 axios 注册为 Vue 插件使用,并将 api 模块挂载在 Vue 原型的 $api 对象上。这样在能获取 this 引用的地方就可以通过 "this.$api.子模块.方法" 的方式调用 API 了。

```
import api from './api'  // 导入所有接口

const install = Vue => {
    if (install.installed)
        return;
    install.installed = true;
    Object.defineProperties(Vue.prototype, {
        // 注意,此处挂载在 Vue 原型的 $api 对象上
        $api: {
            get() {
                return api
            }
        }
    })
}
export default install
```

### 4. api.js

此模块是一个聚合模块,汇合 modules 目录下的所有子模块 API。

```
// 接口统一集成模块
import * as login from './modules/login'
import * as user from './modules/user'
import * as dept from './modules/dept'
import * as role from './modules/role'
import * as menu from './modules/menu'
import * as dict from './modules/dict'
import * as config from './modules/config'
import * as log from './modules/log'
import * as loginlog from './modules/loginlog'
```

```
// 默认全部导出
export default {
    login, user, dept, role, menu, dict, config, log, loginlog
}
```

### 5. user.js

modules 目录下的子模块太多，不方便全贴，这里就以用户管理模块为例。

```
import axios from '../axios'
/* 用户管理模块 */
// 保存
export const save = (data) => {
    return axios({ url: '/user/save', method: 'post', data })
}
// 删除
export const batchDelete = (data) => {
    return axios({ url: '/user/delete', method: 'post', data })
}
// 分页查询
export const findPage = (data) => {
    return axios({ url: '/user/findPage', method: 'post', data })
}
// 查找用户的菜单权限标识集合
export const findPermissions = (params) => {
    return axios({ url: '/user/findPermissions', method: 'get', params })
}
```

### 6. global.js

上面的配置文件中引用了 global.js，我们把一些全局的配置、常量和方法放置在此文件中。

```
/**
 * 全局常量、方法封装模块
 * 通过原型挂载到 Vue 属性
 * 通过 this.global 调用
 */
// 后台管理系统服务器地址
export const baseUrl = 'http://localhost:8001'
// 系统数据备份还原服务器地址
export const backupBaseUrl = 'http://localhost:8002'

export default {
    baseUrl, backupBaseUrl
}
```

### 7. main.js

修改 main.js，导入 API 模块，并通过 Vue.use(api)语句进行使用注册，这样就可以通过"this.$api.子模块.方法"的方式来调用后台接口了。

引入 global 模块，并通过 Vue.prototype.global = global 语句进行挂载，这样就可以通过 this.gloabl.xx 来获取全局配置了。

```
import Vue from 'vue'
import App from './App'
import router from './router'
import api from './http'
import global from '@/utils/global'
import ElementUI from 'element-ui'
import 'element-ui/lib/theme-chalk/index.css'

Vue.use(ElementUI)   // 注册使用 Element
Vue.use(api)     // 注册使用 API 模块

Vue.prototype.global = global // 挂载全局配置模块

new Vue({
  el: '#app',
  router,
  render: h => h(App)
})
```

## 22.1.5 安装 js-cookie

在上面的 axios.js 中，会用到 Cookie 获取 token，所以需要把相关依赖安装一下。执行以下命令，安装依赖包，如图 22-3 所示。

```
yarn add js-cookie
```

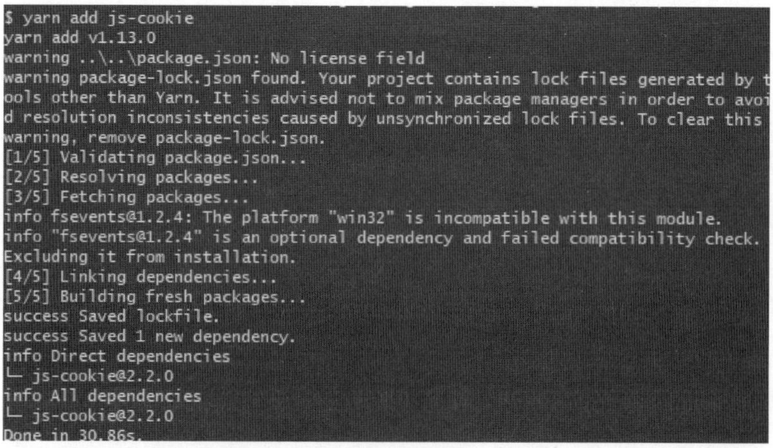

图 22-3

## 22.1.6 测试案例

### 1. 登录页面

在登录界面 Login.vue 中，添加一个登录按钮，单击处理函数通过 axios 调用 login 接口返回数据。

成功返回之后，弹出框，显示 token 信息，然后将 token 放入 Cookie 并跳转到主页。

```
<template>
  <div class="page">
    <h2>Login Page</h2>
    <el-button type="primary" @click="login()">登录</el-button>
  </div>
</template>

<script>
  import mock from '@/mock/mock.js';
  import Cookies from "js-cookie";
  import router from '@/router'
  export default {
    name: 'Login',
    methods: {
      login() {
        this.$api.login.login().then(function(res) {
          alert(res.token)
          Cookies.set('token', res.token) // 放置 token 到 Cookie
          router.push('/')  // 登录成功，跳转到主页
        }).catch(function(res) {
          alert(res);
        });
      }
    }
  }
</script>
```

### 2. Mock 接口

在 mock.js 中添加 login 接口进行拦截，返回一个 token。

```
import Mock from 'mockjs'

Mock.mock('http://localhost:8001/login', {
    'token': '332fr3e3rfsdfd' // 令牌
})
```

```
Mock.mock('http://localhost:8080/user', {
    'name': '@name', // 随机生成姓名
    'name': '@email', // 随机生成邮箱
    'age|1-10': 5, // 年龄在 1~10 岁之间
})
Mock.mock('http://localhost:8080/menu', {
    'id': '@increment', // id自增
    'name': 'menu', // 名称为menu
    'order|1-20': 5, // 排序在 1~20 之间
})
```

3. 页面测试

在浏览器中访问 http://localhost:8080/#/login，显示登录界面，如图 22-4 所示。

图 22-4

单击"登录"按钮，弹出框，显示返回的 token 信息，如图 22-5 所示。

图 22-5

单击"确定"按钮之后，页面跳转到主页，说明我们的 axios 模块已经成功封装并使用了，如图 22-6 所示。

图 22-6

## 22.2 封装 mock 模块

为了可以统一管理和集中控制数据模拟接口，我们对 mock 模块进行了封装，可以方便地定制模拟接口的统一开关和个体开关。

### 22.2.1 文件结构

在 mock 目录下新建一个 index.js，创建 modules 目录并在里面创建子模块的*.js 文件，如图 22-7 所示。

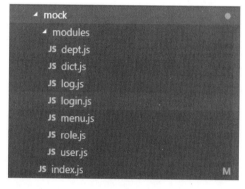

图 22-7

**1. index.js**

index.js 是聚合模块，统一导入所有子模块并通过调用 mock 进行数据模拟。

```
import Mock from 'mockjs'
import { baseUrl } from '@/utils/global'
import * as login from './modules/login'
import * as user from './modules/user'
import * as role from './modules/role'
import * as dept from './modules/dept'
import * as menu from './modules/menu'
import * as dict from './modules/dict'
import * as dict from './modules/dict'
import * as config from './modules/config'
import * as log from './modules/log'
import * as loginlog from './modules/loginlog'

// 1. 开启/关闭[所有模块]拦截，通过调用[openMock 参数]设置
// 2. 开启/关闭[业务模块]拦截，通过调用 fnCreate 方法[isOpen 参数]设置
```

```js
// 3. 开启/关闭[业务模块中某个请求]拦截，通过函数返回对象中的[isOpen 属性]设置
let openMock = true
// let openMock = false
fnCreate(login, openMock)
fnCreate(user, openMock)
fnCreate(role, openMock)
fnCreate(dept, openMock)
fnCreate(menu, openMock)
fnCreate(dict, openMock)
fnCreate(config, openMock)
fnCreate(log, openMock)
fnCreate(loginlog, openMock)

/**
 * 创建mock模拟数据
 * @param {*} mod 模块
 * @param {*} isOpen 是否开启？
 */
function fnCreate (mod, isOpen = true) {

  if (isOpen) {
    for (var key in mod) {
      ((res) => {
        if (res.isOpen !== false) {
          let url = baseUrl
          if(!url.endsWith("/")) {
            url = url + "/"
          }
          url = url + res.url
          Mock.mock(new RegExp(url), res.type, (opts) => {
            opts['data'] = opts.body ? JSON.parse(opts.body) : null
            delete opts.body
            console.log('\n')
            console.log('%cmock拦截，请求: ', 'color:blue', opts)
            console.log('%cmock拦截，响应: ', 'color:blue', res.data)
            return res.data
          })
        }
      })(mod[key]() || {})
    }
  }
}
```

### 2. user.js

子模块 modules 下的代码太多，不方便贴出来，这里以用户管理模块为例，格式跟后台接口保持一致。

```js
/* 用户管理模块 */

// 保存
export function save() {
  return { url: 'user/save', type: 'post',
    data: { "code": 200, "msg": null, "data": 1 }
  }
}
// 批量删除
export function batchDelete() {
  return { url: 'user/delete', type: 'post',
    data: { "code": 200, "msg": null, "data": 1 }
  }
}
// 分页查询
export function findPage(params) {
  let findPageData = { "code": 200, "msg": null, "data": {} }
  let pageNum = 1
  let pageSize = 8
  let content = this.getContent(pageNum, pageSize)
  findPageData.data.pageNum = pageNum
  findPageData.data.pageSize = pageSize
  findPageData.data.totalSize = 50
  findPageData.data.content = content
  return { url: 'user/findPage', type: 'post', data: findPageData }
}
export function getContent(pageNum, pageSize) {
  let content = []
  for(let i=0; i<pageSize; i++) {
    let obj = {}
    let index = ((pageNum - 1) * pageSize) + i + 1
    obj.id = index
    obj.name = 'mango' + index
    obj.password = '9ec9750e709431dad22365cabc5c625482e'
    obj.salt = 'YzcmCZNvbXocrsz9dm8e'
    obj.email = 'mango' + index +'@qq.com'
    obj.mobile = '18688982323'
    obj.status = 1
    obj.deptId = 12
```

```
      obj.deptName = '技术部'
      obj.status = 1
      if(i % 2 === 0) {
        obj.deptId = 13
        obj.deptName = '市场部'
      }
      obj.createBy= 'admin'
      obj.createTime= '2018-08-14 11:11:11'
      obj.createBy= 'admin'
      obj.createTime= '2018-09-14 12:12:12'
      content.push(obj)
  }
  return content
}
// 查找用户的菜单权限标识集合
export function findPermissions() {
  let permsData = { "code": 200, "msg": null,
    "data": [
      "sys:user:view", "sys:menu:delete", "sys:dept:edit", "sys:dict:edit",
"sys:dict:delete", "sys:menu:add","sys:user:add", "sys:log:view",
"sys:dept:delete", "sys:role:edit", "sys:role:view", "sys:dict:view",
      "sys:user:edit", "sys:user:delete", "sys:dept:view","sys:dept:add",
"sys:role:delete", "sys:menu:view","sys:menu:edit", "sys:dict:add"
    ]
  }
  return { url: 'user/findPermissions', type: 'get', data: permsData }
}
```

## 22.2.2 登录界面

修改登录界面，包括导入语句和返回数据格式的获取，如图 22-8 所示。

```
<script>
import mock from '@/mock/index.js'
import Cookies from "js-cookie"
import router from '@/router'
export default {
  name: 'Login',
  methods: {
    login() {
      this.$api.login.login().then(function(res) {
        alert(res.data.token)
        Cookies.set('token', res.data.token) // 放置token到Cookie
        router.push('/')   // 登录成功，跳转到主页
      }).catch(function(res) {
        alert(res);
      });
    }
  }
}
</script>
```

图 22-8

## 22.2.3 主页界面

修改主页界面，替换导入 mock 文件的语句，如图 22-9 所示。

```
<script>
import axios from 'axios'
import mock from '@/mock/index.js'
export default {
  name: 'Home',
  methods: {
    testAxios() {
      axios.get('http://localhost:8080').then(res => { alert(res.data) })
    },
    getUser() {
      axios.get('http://localhost:8080/user').then(res => { alert(JSON.stringify(res.data)) })
    },
    getMenu() {
      axios.get('http://localhost:8080/menu').then(res => { alert(JSON.stringify(res.data)) })
    }
  }
}
</script>
```

图 22-9

## 22.2.4 页面测试

在浏览器中访问 http://localhost:8080/#/login，按照先前的流程走一遍，如果没有问题就封装好了。

在登录界面中，单击"登录"按钮，在弹出框中返回 token 信息，如图 22-10 所示。

图 22-10

登录成功之后，重定向到主页面，如图 22-11 所示。

图 22-11

到此，axios 模块和 mock 模块都封装好了，后续使用只要参考此处的案例即可。

# 第 23 章 第三方图标库

## 23.1 使用第三方图标库

用过 Element 的人都知道，Element UI 提供的字体图符少之又少，实在不够用，幸好现在有不少丰富的第三方图标库可用，引入也不会很麻烦。

## 23.2 Font Awesome

Font Awesome 提供了 675 个可缩放的矢量图标，可以使用 CSS 所提供的所有特性对它们进行更改，包括大小、颜色、阴影或者其他任何支持的效果。

Font Awesome 5 跟之前的版本使用方式差别较大，功能是强大了，图标也更丰富了，但使用也变得更加复杂了。

若是需求没那么复杂，只要简单的有图标可用就行了，还是之前的版本、安装容易、使用简单。

网上相关介绍很多，这里就不废话了，更多详情可参见官方信息，网址为 http://fontawesome.dashgame.com/。

### 23.2.1 安装依赖

执行 yarn add font-awesome 命令，安装 font-awesome 依赖，执行结果如图 23-1 所示。

### 23.2.2 项目引入

在项目 main.js 中引入 css 依赖，可执行 import 'font-awesome/css/font-awesome.min.css'命令，如图 23-2 所示。

```
$ yarn add font-awesome
yarn add v1.13.0
warning package-lock.json found. Your project contains lock files generated by t
ools other than Yarn. It is advised not to mix package managers in order to avoi
d resolution inconsistencies caused by unsynchronized lock files. To clear this
warning, remove package-lock.json.
[1/5] Validating package.json...
[2/5] Resolving packages...
[3/5] Fetching packages...
info fsevents@1.2.4: The platform "win32" is incompatible with this module.
info "fsevents@1.2.4" is an optional dependency and failed compatibility check.
Excluding it from installation.
[4/5] Linking dependencies...
[5/5] Building fresh packages...
success Saved 1 new dependency.
info Direct dependencies
└─ font-awesome@4.7.0
info All dependencies
└─ font-awesome@4.7.0
Done in 26.50s.
```

图 23-1

```
import 'element-ui/lib/theme-chalk/index.css'
import 'font-awesome/css/font-awesome.min.css'

Vue.use(ElementUI)  // 注册使用Element
Vue.use(api)  // 注册使用API模块

Vue.prototype.global = global // 挂载全局配置模块
```

图 23-2

## 23.2.3　页面使用

项目引入之后，直接在页面使用就可以了，修改 Home.vue，加入一个图标，如图 23-3 所示。

```
<template>
  <div class="page">
    <h2>Home Page</h2>
    <li class="fa fa-home fa-lg"></li>
    <el-button type="primary" @click="testAxios()">测试Axios调用</el-button>
    <el-button type="primary" @click="getUser()">获取用户信息</el-button>
    <el-button type="primary" @click="getMenu()">获取菜单信息</el-button>
  </div>
</template>
```

图 23-3

## 23.2.4　页面测试

启动应用，访问 http://localhost:8080/#/，效果如图 23-4 所示。

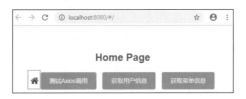

图 23-4

就是这么简单，就是这么好用！

另外，还可以选择 CDN 方式、下载方式等，这里就不说了，有兴趣的可自行查阅。

- 官方网址：http://fontawesome.dashgame.com/
- 官方教程：https://fontawesome.com/how-to-use

# 第 24 章 多语言国际化

国际化多语言支持是现在系统通常都要具备的功能，Vue 对国际化提供了很好的支持，本章我们就来讲解如何实现国际化。

## 24.1 安装依赖

首先需要安装国际化组件，执行 yarn add vue-i18n 命令，安装 i18n 依赖，如图 24-1 所示。

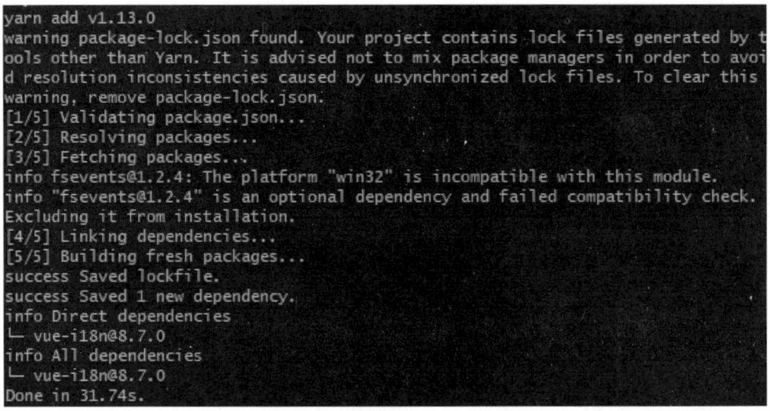

```
yarn add v1.13.0
warning package-lock.json found. Your project contains lock files generated by t
ools other than Yarn. It is advised not to mix package managers in order to avoi
d resolution inconsistencies caused by unsynchronized lock files. To clear this
warning, remove package-lock.json.
[1/5] Validating package.json...
[2/5] Resolving packages...
[3/5] Fetching packages...
info fsevents@1.2.4: The platform "win32" is incompatible with this module.
info "fsevents@1.2.4" is an optional dependency and failed compatibility check.
Excluding it from installation.
[4/5] Linking dependencies...
[5/5] Building fresh packages...
success Saved lockfile.
success Saved 1 new dependency.
info Direct dependencies
└─ vue-i18n@8.7.0
info All dependencies
└─ vue-i18n@8.7.0
Done in 31.74s.
```

图 24-1

## 24.2 添加配置

在 src 下新建 i18n 目录，并创建一个 index.js。

```
import Vue from 'vue'
import VueI18n from 'vue-i18n'

Vue.use(VueI18n)

// 注册 i18n 实例并引入语言文件，文件格式等一下解析
```

```
const i18n = new VueI18n({
  locale: 'zh_cn',
  messages: {
    'zh_cn': require('@/assets/languages/zh_cn.json'),
    'en_us': require('@/assets/languages/en_us.json')
  }
})

export default i18n
```

然后在 assets 目录下面创建两个多语言文件,如图 24-2 所示。

图 24-2

zh_cn.json

```
{
    "common": {
        "home":"首页",
        "login":"登录",
        "logout":"退出登录",
        "doc":"文档",
        "blog":"博客",
        // 省略其他
    }
}
```

en_us.json

```
{
    "common": {
        "home": "Home",
        "login": "Login",
        "logout": "Logout",
        "doc": "Document",
        "blog": "Blog",
        // 省略其他
    }
}
```

在 main.js 中引入 i18n 并注入 vue 对象中。

```
import Vue from 'vue'
import App from './App'
import router from './router'
import api from './http'
import i18n from './i18n'
import global from '@/utils/global'
import ElementUI from 'element-ui'
import 'element-ui/lib/theme-chalk/index.css'
import 'font-awesome/css/font-awesome.min.css'

Vue.use(ElementUI)   // 注册使用 Element
Vue.use(api)    // 注册使用 API 模块

Vue.prototype.global = global // 挂载全局配置模块

new Vue({
  el: '#app',
  i18n,
  router,
  render: h => h(App)
})
```

## 24.3 字符引用

在原本使用字符串的地方引入国际化字符串。

打开 Home.vue，在模板下面添加一个国际化字符串和两个按钮做中英文切换。

```
    <h3>{{$t('common.doc')}}</h3>
    <el-button type="success" @click="changeLanguage('zh_cn')">简体中文</el-button>
    <el-button type="success" @click="changeLanguage('en_us')">English</el-button>
```

在方法声明区域添加以下方法，设置国际化语言。

```
    // 语言切换
    changeLanguage(lang) {
      lang === '' ? 'zh_cn' : lang
      this.$i18n.locale = lang
      this.langVisible = false
    }
```

## 24.4 页面测试

启动应用，访问 http://localhost:8080/#/，进入主页，如图 24-3 所示。

图 24-3

单击 English 按钮，国际化字符变成英文，如图 24-4 所示。

图 24-4

单击"简体中文"按钮，国际化字符变回中文，如图 24-5 所示。

图 24-5

通过 this.$i18n.locale = xx 方式就可以全局切换语言，Vue 框架会根据 locale 的值读取对应的国际化多语言文件并进行适时更新。

# 第 25 章 登录流程完善

## 25.1 登录界面

### 25.1.1 界面设计

首先我们简单设计一下登录界面，标清楚就行，然后照着写页面，如图 25-1 所示。

图 25-1

### 25.1.2 关键代码

由于页面内容太多，不便全贴，至于页面组件布局什么的直接查阅代码即可。这里重点讲解一下登录的逻辑。登录方法内容如下面所示，主要逻辑是调用后台登录接口，并在登录成功之后保存 token 到 Cookie，保存用户到本地存储，然后跳转到主页面。

```
login() {
    this.loading = true
    let userInfo = { account:this.loginForm.account, password:this.loginForm.password,
        captcha:this.loginForm.captcha }
    this.$api.login.login(userInfo).then((res) => {  // 调用登录接口
        if(res.msg != null) {
            this.$message({ message: res.msg, type: 'error' })
```

```
      } else {
        Cookies.set('token', res.data.token) // 放置 token 到 Cookie
        sessionStorage.setItem('user', userInfo.account) // 保存用户到本地会话
        this.$router.push('/')  // 登录成功，跳转到主页
      }
      this.loading = false
    }).catch((res) => {
      this.$message({ message: res.message, type: 'error' })
    })
  }
```

## 25.2 主页面

### 25.2.1 界面设计

先简单设计一下主页面整体框架，再照着设计图编写页面，如图 25-2 所示。

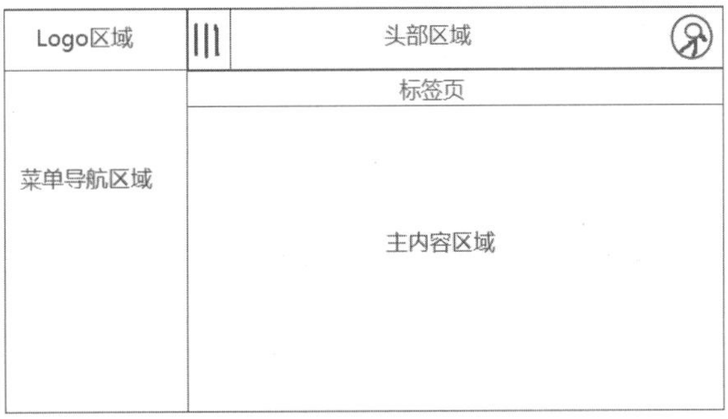

图 25-2

### 25.2.2 关键代码

页面内容太多，不便全贴，至于页面组件布局什么的请读者直接查阅代码即可。我们在 views 下面另外添加几个页面文件，分别为头部、左侧导航和主内容区域，如图 25-3 所示。

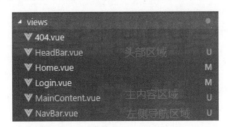

图 25-3

### 1. Home.vue

主页由导航菜单、头部区域和主内容区域组成。

```html
<template>
  <div class="container">
     <!-- 导航菜单栏 -->
     <nav-bar></nav-bar>
     <!-- 头部区域 -->
     <head-bar></head-bar>
     <!-- 主内容区域 -->
     <main-content></main-content>
  </div>
</template>

<script>
import HeadBar from "./HeadBar"
import NavBar from "./NavBar"
import MainContent from "./MainContent"
export default {
  components:{
      HeadBar,
      NavBar,
      MainContent
  }
};
</script>

// css 代码省略
```

### 2. HeadBar.vue

头部导航主要是设置样式，并在右侧添加用户名和头像显示。

```html
<template>
  <div class="headbar" style="background:#14889A" :class="'position-left'">
   <!-- 工具栏 -->
   <span class="toolbar">
      <el-menu class="el-menu-demo" background-color="#14889A"
text-color="#14889A" active-text-color="#14889A" mode="horizontal">
         <el-menu-item index="1">
           <!-- 用户信息 -->
           <span class="user-info"><img :src="user.avatar" />{{user.name}}</span>
         </el-menu-item>
```

```
      </el-menu>
    </span>
  </div>
</template>

// 省略 js 和 css 代码
```

### 3. NavBar.vue

左侧导航包含上方 Logo 区域和下方导航菜单区域。

```
<template>
  <div class="menu-bar-container">
    <!-- logo -->
    <div class="logo" style="background:#14889A" :class="'menu-bar-width'"
        @click="$router.push('/')">
        <img src="@/assets/logo.png"/> <div>Mango</div>
    </div>
  </div>
</template>
```

// 省略 css 样式

### 4. MainContent.vue

主内容区域包含标签页导航和主内容区域,在主内容中放置 route-view,用于路由信息。

```
<template>
  <div id="main-container" class="main-container" :class="'position-left'">
    <!-- 标签页 -->
    <div class="tab-container"></div>
    <!-- 主内容区域 -->
    <div class="main-content">
      <keep-alive>
        <transition name="fade" mode="out-in">
          <router-view></router-view>
        </transition>
      </keep-alive>
    </div>
  </div>
</template>

// 省略 css 样式
```

## 25.3 页面测试

启动应用，访问 http://localhost:8080/#/login，进入登录界面，如图 25-4 所示。

图 25-4

单击"登录"按钮，登录成功之后跳转到主页面，如图 25-5 所示。

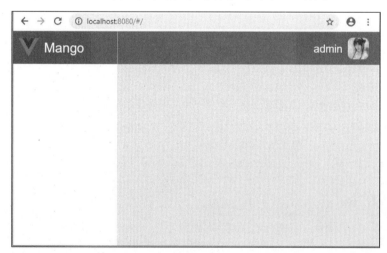

图 25-5

# 第 26 章 管理应用状态

在很多应用场景下，我们需要在组件之间共享状态，比如我们的左侧导航栏需要收缩和展开的功能，收缩状态时宽度很小，只显示菜单图标，因为导航菜单栏收缩之后宽度变了，所以右侧的主内容区域要占用导航栏收缩的空间，主内容区域宽度也要根据导航栏的收缩状态做变更，而导航栏和主内容区域是两个不同的组件，而非父子组件之间不支持状态传递，所以组件之间的状态共享问题发生了。vuex 是一个专为 vue.js 应用程序开发的状态管理模式。它采用集中式存储管理应用的所有组件的状态，并以相应的规则保证状态以一种可预测的方式发生变化。本章将通过实现左侧导航栏的收缩展开功能来讲解怎样使用 vuex 来管理应用状态。

更多 vuex 的资料可参考：https://vuex.vuejs.org/zh/。

## 26.1 安装依赖

执行 yarn add vuex 命令，安装 vuex 依赖，如图 26-1 所示。

```
$ yarn add vuex
yarn add v1.13.0
warning package-lock.json found. Your project contains lock files generated by t
ools other than Yarn. It is advised not to mix package managers in order to avoi
d resolution inconsistencies caused by unsynchronized lock files. To clear this
warning, remove package-lock.json.
[1/5] Validating package.json...
[2/5] Resolving packages...
[3/5] Fetching packages...
info fsevents@1.2.4: The platform "win32" is incompatible with this module.
info "fsevents@1.2.4" is an optional dependency and failed compatibility check.
Excluding it from installation.
[4/5] Linking dependencies...
[5/5] Building fresh packages...
success Saved lockfile.
success Saved 1 new dependency.
info Direct dependencies
└─ vuex@3.0.1
info All dependencies
└─ vuex@3.0.1
Done in 29.91s.
```

图 26-1

## 26.2 添加 store

在 src 目录下新建一个 store 目录，专门管理应用状态，如图 26-2 所示。

图 26-2

### 26.2.1 index.js

在 index.js 中引入 vuex 并统一组织导入和管理子模块。

```
import Vue from 'vue'
import vuex from 'vuex'

Vue.use(vuex);

// 引入子模块
import app from './modules/app'

const store = new vuex.Store({
    modules: {
        app: app
    }
})

export default store
```

### 26.2.2 app.js

app.js 是属于应用内的全局性的配置，比如主题色、导航栏收缩状态等，详见注释。

```
export default {
    state: {
        appName: "Mango Platform",   // 应用名称
        themeColor: "#14889A",   // 主题颜色
        oldThemeColor: "#14889A",    // 上一次主题颜色
        collapse:false,   // 导航栏收缩状态
        menuRouteLoaded:false      // 菜单和路由是否已经加载
    },
    getters: {
        collapse(state){// 对应着上面的 state
            return state.collapse
        }
    },
    mutations: {
        onCollapse(state){   // 改变收缩状态
```

```
            state.collapse = !state.collapse
        },
        setThemeColor(state, themeColor){   // 改变主题颜色
            state.oldThemeColor = state.themeColor
            state.themeColor = themeColor
        },
        menuRouteLoaded(state, menuRouteLoaded){   // 改变菜单和路由的加载状态
            state.menuRouteLoaded = menuRouteLoaded;
        }
    },
    actions: {
    }
}
```

## 26.3 引入 Store

在 main.js 中引入 store，如图 26-3 所示。

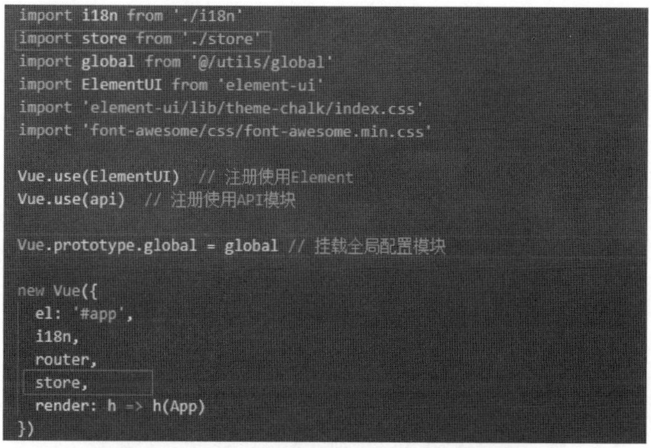

```
import i18n from './i18n'
import store from './store'
import global from '@/utils/global'
import ElementUI from 'element-ui'
import 'element-ui/lib/theme-chalk/index.css'
import 'font-awesome/css/font-awesome.min.css'

Vue.use(ElementUI)   // 注册使用Element
Vue.use(api)   // 注册使用API模块

Vue.prototype.global = global   // 挂载全局配置模块

new Vue({
    el: '#app',
    i18n,
    router,
    store,
    render: h => h(App)
})
```

图 26-3

## 26.4 使用 Store

这里以头部页面 HeadBar.vue 的状态使用为例，其他页面同理，详见代码。

首先通过 computed 计算属性引入 store 属性，这样就可以直接在页面中通过 collapse 引用状态值了，当然如果不嫌长，也可以不使用计算属性，直接在页面中通过$store.state.app.collapse 引用，如图 26-4 所示。

图 26-4

然后在页面中通过 collapse 的状态值来绑定不同的宽度样式，如图 26-5 所示。

图 26-5

## 26.5 收缩组件

在 src 下新建 components 目录，并在其下创建导航栏收缩展开组件 Hamburger。

### 26.5.1 文件结构

文件结构如图 26-6 所示。

图 26-6

### 26.5.2 关键代码

组件是使用 SVG 绘制，绘制根据 isActive 状态决定是否旋转、显示收缩和展开状态不同的图形，如图 26-7 所示。

在头部区域 HeadBar 中引入 hamburger，并将自身 isActive 状态跟收缩状态 collapse 绑定，如图 26-8 所示。

图 26-7

图 26-8

单击导航栏收缩组件区域的响应函数，设置导航收缩状态到 Store。

```
// 折叠导航栏
onCollapse: function() {
  this.$store.commit('onCollapse')
}
```

# 26.6 页面测试

启动应用，访问 http://localhost:8080/#/login，单击"登录"按钮进入主页面，如图 26-9 所示。

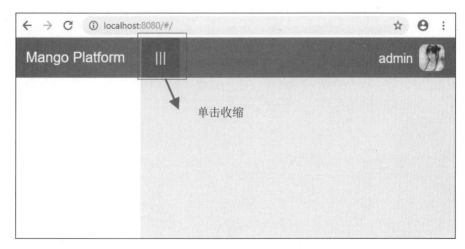

图 26-9

单击导航栏收缩展开组件,导航菜单栏收缩起来,效果如图 26-10 所示。

图 26-10

# 第 27 章 头部功能组件

本章我们来介绍头部区域一些常用功能的实现方案，比如动态主题切换器、国际化语言切换器、用户信息弹出面板等。

## 27.1 主题切换组件

### 27.1.1 编写组件

在 components 目录下创建一个主题切换器组件 ThemePicker，如图 27-1 所示。

图 27-1

ThemePicker 的实现思路是使用一个颜色选取组件 el-color-picker 获取一个主题色，然后通过动态替换覆盖 Element 默认 CSS 样式的方式替换框架的主题色 primary color，并在主题色切换成功之后提供回调函数，通过此回调函数同步更新需要更换为主题色的页面或组件。组件代码篇幅太多不便全贴，下面讲解一下关键代码。

组件使用 el-color-picker 获取主题色，并绑定 theme 属性和 size 属性。

```
<template>
 <el-color-picker class="theme-picker" popper-class="theme-picker-dropdown"
  v-model="theme" :size="size">
 </el-color-picker>
</template>
```

通过 watch 监听 theme 属性即主题色的更新动态替换 CSS 样式，修改主题色，并在替换 CSS 之后通过 this.$emit('onThemeChange', val)语句提供一个回调函数'onThemeChange'并将更新后的主题色的值 val 作为参数传入，使得外部组件可以通过此回调函数同步更新外部组件颜色为主题色。

```js
watch: {
  theme(val, oldVal) {
    if (typeof val !== 'string') return
    // 替换 CSS 样式，修改主题色
    const themeCluster = this.getThemeCluster(val.replace('#', ''))
    const originalCluster = this.getThemeCluster(oldVal.replace('#', ''))
    console.log(themeCluster, originalCluster)
    const getHandler = (variable, id) => {
      return () => {
        const originalCluster = this.getThemeCluster(ORIGINAL_THEME.replace('#', ''))
        const newStyle = this.updateStyle(this[variable], originalCluster, themeCluster)

        let styleTag = document.getElementById(id)
        if (!styleTag) {
          styleTag = document.createElement('style')
          styleTag.setAttribute('id', id)
          document.head.appendChild(styleTag)
        }
        styleTag.innerText = newStyle
      }
    }
    const chalkHandler = getHandler('chalk', 'chalk-style')
    if (!this.chalk) {
      const url = `https://unpkg.com/element-ui@${version}/lib/theme-chalk/index.css`
      this.getCSSString(url, chalkHandler, 'chalk')
    } else {
      chalkHandler()
    }
    const styles = [].slice.call(document.querySelectorAll('style'))
      .filter(style => {
        const text = style.innerText
        return new RegExp(oldVal, 'i').test(text) && !/Chalk Variables/.test(text)
      })
    styles.forEach(style => {
      const { innerText } = style
      if (typeof innerText !== 'string') return
      style.innerText = this.updateStyle(innerText, originalCluster, themeCluster)
    })
```

```
    // 响应外部操作
    this.$emit('onThemeChange', val)
    if(this.showSuccess) {
      this.$message({ message: '换肤成功', type: 'success' })
    } else {
      this.showSuccess = true
    }
  }
}
```

在 HeadBar.vue 中引入 ThemePicker 组件，如图 27-2 所示。

```
import Hamburger from "@/components/Hamburger"
import ThemePicker from "@/components/ThemePicker"
export default {
  components:{
      Hamburger,
      ThemePicker
  },
```

图 27-2

在页面中添加一个皮肤主题色切换器菜单，如图 27-3 所示。

```
<!-- 工具栏 -->
<span class="toolbar">
  <el-menu class="el-menu-demo" :background-color="themeColor" text-color="#14889A"
    :active-text-color="themeColor" mode="horizontal">
    <el-menu-item index="1">
      <!-- 主题切换 -->
      <theme-picker class="theme-picker" :default="themeColor"
        @onThemeChange="onThemeChange">
      </theme-picker>
    </el-menu-item>
    <el-menu-item index="2">
      <!-- 用户信息 -->
      <span class="user-info"><img :src="user.avatar" />{{user.name}}</span>
    </el-menu-item>
```

图 27-3

在方法区定义主题切换的回调函数，同步设置 store 中的 themeColor，如图 27-4 所示。

```
// 折叠导航栏
onCollapse: function() {
  this.$store.commit('onCollapse')
},
// 切换主题
onThemeChange: function(themeColor) {
  this.$store.commit('setThemeColor', themeColor)
},
```

图 27-4

将各个页面中需要同步为主题色的页面或组件绑定 store 中的主题色属性 themeColor，这样每次通过切换器切换主题色的时候都会把主题色通过 store 传递到页面组件，如图 27-5 所示。

```
<template>
  <div class="headbar" :style="{'background':themeColor}"
    :class="collapse?'position-collapse-left':'position-left'">
    <!-- 导航收缩 -->
    <span class="hamburg">
      <el-menu class="el-menu-demo" :background-color="themeColor" text-color="#fff"
        :active-text-color="themeColor" mode="horizontal">
        <el-menu-item index="1" @click="onCollapse">
          <hamburger :isActive="collapse"></hamburger>
        </el-menu-item>
      </el-menu>
    </span>
```

图 27-5

### 27.1.2 页面测试

启动应用，访问 http://localhost:8080/#/，单击图 27-6 所示的切换按钮，选择一个主题色。

图 27-6

选取一个主题色之后，单击"确定"按钮，页面主题色成功替换，如图 27-7 所示。

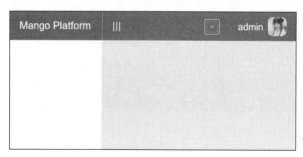

图 27-7

## 27.2 语言切换组件

### 27.2.1 编写组件

在 HeadBar.vue 工具栏主题切换器右边放置一个语言切换组件，可以选择中英文两种语言，如图 27-8 所示。

语言切换响应函数，设置 local 值并让弹出下拉面板消失，如图 27-9 所示。

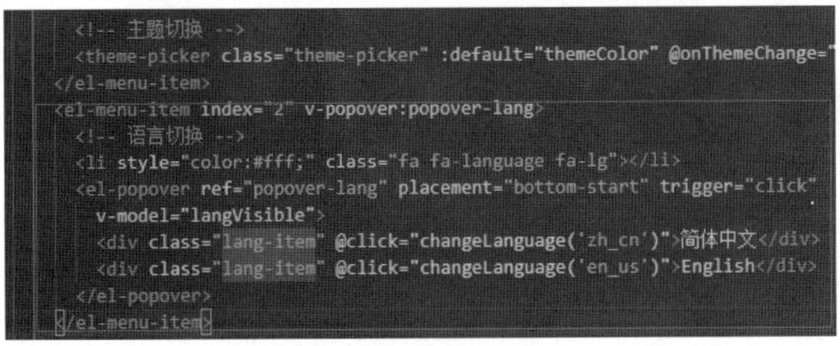

图 27-8 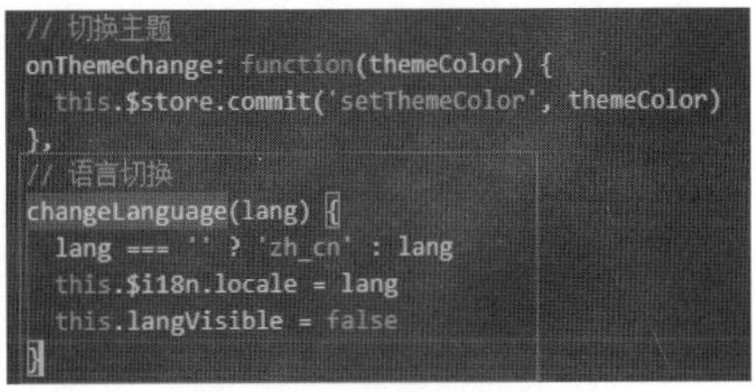　　　　　　　　　　　　　图 27-9

在头部工具栏左侧放置一些菜单项，使用国际化字符串，以测试多语言切换效果，如图 27-10 所示。

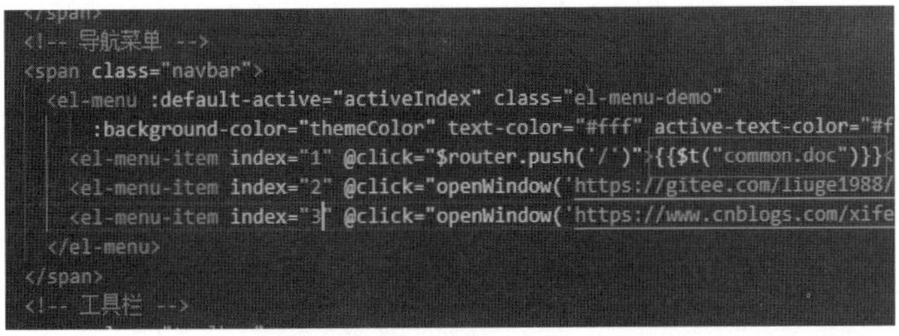

图 27-10

## 27.2.2　页面测试

启动应用，访问 http://localhost:8080/#/，单击如图 27-11 所示的语言切换按钮，选择语言为英文。

图 27-11

页面上的国际化字符串成功切换为英文字符，如图 27-12 所示。

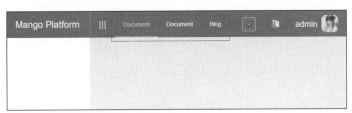

图 27-12

## 27.3 用户信息面板

### 27.3.1 编写组件

在 views 目录下新建 core 目录并在其下新建一个用户信息面板 PersonalPanel.vue，在单击用户头像时弹出用户信息面板，面板显示用户信息和一些功能操作。

用户信息面板页面内容如下：

```
<template>
  <div class="personal-panel">
<div class="personal-desc" :style="{'background':
this.$store.state.app.themeColor}">
      <div class="avatar-container">
        <img class="avatar" :src="require('@/assets/user.png')" />
      </div>
      <div class="name-role">
        <span class="sender">{{ user.name }} - {{ user.role }}</span>
      </div>
      <div class="registe-info">
        <span class="registe-info"> <li class="fa fa-clock-o"></li>
          {{ user.registeInfo }}
        </span>
      </div>
    </div>
    <div class="personal-relation">
      <span class="relation-item">followers</span>
      <span class="relation-item">watches</span>
      <span class="relation-item">friends</span>
    </div>
    <div class="main-operation">
      <span class="main-operation-item">
        <el-button size="small" icon="fa fa-male"> 个人中心</el-button>
      </span>
      <span class="main-operation-item">
        <el-button size="small" icon="fa fa-key"> 修改密码</el-button>
      </span>
    </div>
    <div class="other-operation">
      <div class="other-operation-item">
        <li class="fa fa-eraser"></li> 清除缓存</div>
```

```html
        <div class="other-operation-item">
          <li class="fa fa-user"></li> 在线人数</div>
        <div class="other-operation-item">
          <li class="fa fa-bell"></li> 访问次数</div>
        <div class="other-operation-item">
          <li class="fa fa-undo"></li>{{$t("common.backupRestore")}}</div>
      </div>
      <div class="personal-footer" @click="logout">
        <li class="fa fa-sign-out"></li> {{$t("common.logout")}}
      </div>
    </div>
</template>
```
// 省略其他内容

退出登录处理函数，确认退出后清空本地存储，返回登录界面并调用后台退出登录接口，如图 27-13 所示。

```js
// 退出登录
logout: function() {
  this.$confirm("确认退出吗?", "提示", {
    type: "warning"
  })
  .then(() => {
    sessionStorage.removeItem("user")
    this.$router.push("/login")
    this.$api.login.logout().then((res) => {
    }).catch(function(res) {
    })
  })
  .catch(() => {})
}
```

图 27-13

在头部区域引入组件，在用户头像信息组件下通过 popover 组件关联用户信息面板，如图 27-14 所示。

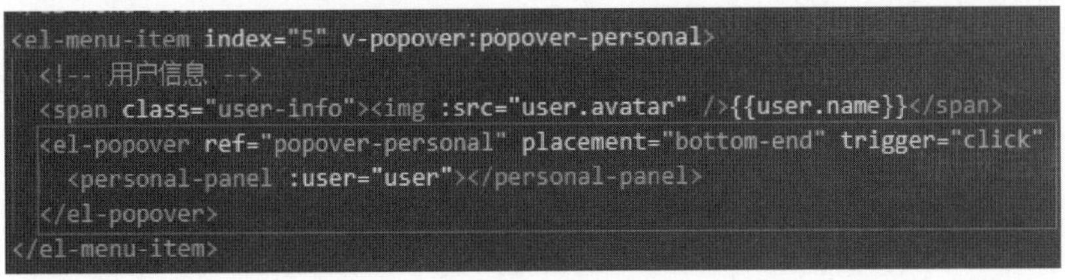

```html
<el-menu-item index="5" v-popover:popover-personal>
  <!-- 用户信息 -->
  <span class="user-info"><img :src="user.avatar" />{{user.name}}</span>
  <el-popover ref="popover-personal" placement="bottom-end" trigger="click">
    <personal-panel :user="user"></personal-panel>
  </el-popover>
</el-menu-item>
```

图 27-14

## 27.3.2 页面测试

启动应用，访问 http://localhost:8080/#/login，单击"登录"按钮，进入主页面，单击头像区域，弹出用户信息面板，效果如图 27-15 所示。

图 27-15

单击信息面板最下面的"退出登录"按钮，弹出退出确认提示框，单击"确定"按钮返回登录界面，如图 27-16 所示。

图 27-16

## 27.4 系统通知面板

### 27.4.1 编写组件

在 views/core 目录下新建一个系统通知面板 NoticePanel.vue，在头部工具栏添加系统通知组件，单击弹出系统通知信息面板。

系统通知面板页面内容如下，主要通过遍历通知信息列表展示。

```
<template>
  <div class="notice-panel">
    <div class="header">您有 {{data.length}} 条通知</div>
    <div class="notice-content">
      <div v-for="item in data" :key="item.key" class="notice-item">
        <span class="notice-icon">
          <li :class="item.icon"></li>
        </span>
        <span class="notice-cotent">
          {{ item.content }}
```

```
          </span>
        </div>
      </div>
      <div class="notice-footer">查看所有通知</div>
    </div>
</template>
```

// 省略其他信息

在头部区域 HeadBar.vue 中引入组件并通过 popover 组件关联通知面板，如图 27-17 所示。

```
<el-menu-item index="4" v-popover:popover-notice>
  <!-- 系统通知 -->
  <el-badge :value="4" :max="99" class="badge" type="error">
    <li style="color:#fff;" class="fa fa-bell-o fa-lg"></li>
  </el-badge>
  <el-popover ref="popover-notice" placement="bottom-end" trigger="click">
    <notice-panel></notice-panel>
  </el-popover>
</el-menu-item>
```

图 27-17

## 27.4.2 页面测试

启动应用，访问 http://localhost:8080/#/login，单击"登录"按钮，进入主页面，单击通知区域，弹出系统通知面板，效果如图 27-18 所示。

图 27-18

# 27.5 用户私信面板

## 27.5.1 编写组件

在 views/core 目录下新建一个用户私信面板 MessagePanel.vue，在头部工具栏添加用户私信组件，单击弹出用户私信信息面板。

用户私信面板页面内容如下，主要通过遍历私信信息列表展示。

```html
<template>
  <div class="message-panel">
    <div class="message-header">您有 {{data.length}} 条消息</div>
    <div class="message-content">
      <div v-for="item in data" :key="item.key" class="message-item">
        <div class="message-avatar">
          <img class="avatar" :src="require('@/assets/user.png')" />
        </div>
        <span class="sender">
          {{ item.sender }}
        </span>
        <span class="time">
          <li class="fa fa-clock-o"></li> {{ item.time }}
        </span>
        <div class="message-cotent">
          {{ item.content }}
        </div>
      </div>
    </div>
    <div class="message-footer">查看所有消息</div>
  </div>
</template>
```

在头部区域 HeadBar.vue 中引入组件并通过 popover 组件关联私信面板，如图 27-19 所示。

```html
<el-menu-item index="3" v-popover:popover-message>
  <!-- 我的私信 -->
  <el-badge :value="5" :max="99" class="badge" type="error">
    <li style="color:#fff;" class="fa fa-envelope-o fa-lg"></li>
  </el-badge>
  <el-popover ref="popover-message" placement="bottom-end" trigger="click"
    <message-panel></message-panel>
  </el-popover>
</el-menu-item>
```

图 27-19

## 27.5.2 页面测试

启动应用，访问 http://localhost:8080/#/login，单击"登录"按钮，进入主页面，单击私信区域，弹出用户私信面板，效果如图 27-20 所示。

第 27 章 头部功能组件

图 27-20

# 第 28 章 动态加载菜单

本章我们将讲解如何动态加载数据库的菜单数据并显示到导航栏。

## 28.1 添加 Store

我们先添加几个 store 状态，后续需要用来共享使用。

首先在 store/modules 下的 app.js 中添加一个 menuRouteLoaded 状态，判断路由是否加载过，如图 28-1 所示。

```
state: {
    appName: "Kitty Platform",    // 应用名称
    themeColor: "#14889A",        // 主题颜色
    oldThemeColor: "#14889A",     // 上一次主题颜色
    collapse:false,               // 导航栏收缩状态
    menuRouteLoaded:false         // 菜单和路由是否已经加载
},
```

图 28-1

然后在 store/modules 下新建一个 menu.js，在 index.js 中引入，里面保存着加载后的导航菜单树数据。

```
export default {
   state: {
      navTree: [],  // 导航菜单树
   },
   getters: {

   },
   mutations: {
      setNavTree(state, navTree){  // 设置导航菜单树
         state.navTree = navTree;
      }
   }
}
```

在 store/modules 下新建一个 user.js，在 index.js 中引入，里面保存加载后的用户权限数据。

```
export default {
    state: {
        perms: [],  // 用户权限标识集合
    },
    getters: {

    },
    mutations: {
        setPerms(state, perms){   // 用户权限标识集合
            state.perms = perms;
        }
    }
}
```

## 28.2 登录页面

打开登录页面 Login.vue，在登录接口设置菜单加载状态，要求重新登录之后重新加载菜单，如图 28-2 所示。

```
this.$api.login.login(userInfo).then((res) => {   // 调用登录接口
    if(res.msg != null) {
        this.$message({ message: res.msg, type: 'error' })
    } else {
        Cookies.set('token', res.data.token) // 放置token到Cookie
        sessionStorage.setItem('user', userInfo.account) // 保存用户到本地会话
        this.$store.commit('menuRouteLoaded', false)  // 要求重新加载导航菜单
        this.$router.push('/')  // 登录成功，跳转到主页
    }
    this.loading = false
}).catch((res) => {
```

图 28-2

## 28.3 导航守卫

路由对象 router 给我们提供了 beforeEach 方法，可以在每次路由之前进行一些相关处理，也叫导航守卫，我们这里就通过导航守卫实现动态菜单的加载。

修改 router/index.js 文件，添加导航守卫，在每次路由时判断用户会话是否过期。如果登录有效且跳转的是登录界面，就直接路由到主页；如果是非登录页面且会话过期，就跳到登录页面要求登录；否则，加载动态菜单和路由并路由到目标页面。

```
router.beforeEach((to, from, next) => {
    // 登录界面登录成功之后，会把用户信息保存在会话
```

```
// 存在时间为会话生命周期，页面关闭即失效
let userName = sessionStorage.getItem('user')
if (to.path === '/login') {
  // 如果是访问登录界面，且用户会话信息存在，则代表已登录过，跳转到主页
  if(userName) {
    next({ path: '/' })
  } else {
    next()
  }
} else {
  if (!userName) {
    // 如果访问非登录界面，且用户会话信息不存在，则代表未登录，跳转到登录界面
    next({ path: '/login' })
  } else {
    // 加载动态菜单和路由
    addDynamicMenuAndRoutes(userName, to, from)
    next()
  }
}
})
```

加载动态路由的方法内容如下。首先判断动态菜单是否已经存在，如果存在就不再重复加载损耗性能，否则调用后台接口加载数据库存储菜单数据，加载成功之后通过 router.addRoutes 方法将菜单数据动态添加到路由器并同时保存菜单数据及加载状态以备后用。导航菜单加载成功之后，调用后台接口查找用户权限数据并保存起来，供权限判断时读取。

```
/**
 * 加载动态菜单和路由
 */
function addDynamicMenuAndRoutes(userName, to, from) {
  if(store.state.app.menuRouteLoaded) {
    console.log('动态菜单和路由已经存在.')
    return
  }
  api.menu.findNavTree({'userName':userName})
  .then(res => {
    // 添加动态路由
    let dynamicRoutes = addDynamicRoutes(res.data)
    router.options.routes[0].children =
router.options.routes[0].children.concat(dynamicRoutes)
    router.addRoutes(router.options.routes)
    // 保存加载状态
    store.commit('menuRouteLoaded', true)
    // 保存菜单树
```

```js
      store.commit('setNavTree', res.data)
    }).then(res => {
      api.user.findPermissions({'name':userName}).then(res => {
        // 保存用户权限标识集合
        store.commit('setPerms', res.data)
      })
    })
    .catch(function(res) {
    })
}
```

下面是遍历菜单数据实际创建路由对象的逻辑。

```js
/**
 * 添加动态(菜单)路由
 * @param {*} menuList 菜单列表
 * @param {*} routes 递归创建的动态(菜单)路由
 */
function addDynamicRoutes (menuList = [], routes = []) {
  var temp = []
  for (var i = 0; i < menuList.length; i++) {
    if (menuList[i].children && menuList[i].children.length >= 1) {
      temp = temp.concat(menuList[i].children)
    } else if (menuList[i].url && /\S/.test(menuList[i].url)) {
      menuList[i].url = menuList[i].url.replace(/^\//, '')
      // 创建路由配置
      var route = {
        path: menuList[i].url, component: null, name: menuList[i].name,
        meta: { icon: menuList[i].icon, index: menuList[i].id }
      }
      try {
        // 根据菜单 URL 动态加载 vue 组件，这里要求 vue 组件须按照 url 路径存储
        // 若 url="sys/user"，则组件路径应是"@/views/sys/user.vue",否则组件加载不到
        let array = menuList[i].url.split('/')
        let url = ''
        for(let i=0; i<array.length; i++) {
          url+=array[i].substring(0,1).toUpperCase()+array[i].substring(1)+'/'
        }
        url = url.substring(0, url.length - 1)
        route['component'] = resolve => require([`@/views/${url}`], resolve)
      } catch (e) {}
      routes.push(route)
    }
  }
}
```

```
    if (temp.length >= 1) {
      addDynamicRoutes(temp, routes)
    }
    return routes
}
```

## 28.4 导航树组件

在 components 目录下新建一个导航树组件 MenuTree，文件结构如图 28-3 所示。

图 28-3

MenuTree.vue 的页面内容大致如下，主要是遍历菜单数据创建导航菜单。

```
<template>
  <el-submenu v-if="menu.children && menu.children.length >= 1" :index="'' + menu.id">
    <template slot="title">
      <i :class="menu.icon" ></i>
      <span slot="title">{{menu.name}}</span>
    </template>
    <MenuTree v-for="item in menu.children" :key="item.id" :menu="item"></MenuTree>
  </el-submenu>
  <el-menu-item v-else :index="'' + menu.id" @click="handleRoute(menu)">
    <i :class="menu.icon"></i>
    <span slot="title">{{menu.name}}</span>
  </el-menu-item>
</template>
```

上面的导航菜单项都绑定了 handleRoute 函数，在单击菜单项的时候路由到指定路径。

路由业务功能页面目前还没有实现，后面实现业务功能页面时会讲到，如图 28-4 所示。

```
methods: {
  handleRoute (menu) {
    // 通过菜单URL跳转至指定路由
    this.$router.push("/" + menu.url)
  }
}
```

图 28-4

在头部区域 HeadBar.vue 中引入导航菜单树组件，并编写导航菜单区域内容，其中的菜单数据 navTree 是在导航守卫加载并存储到 store 的，如图 28-5 所示。

```html
<div class="menu-bar-container">
  <!-- logo -->
  <div class="logo" :style="{'background':themeColor}" @click="$router.push('/')"
    :class="collapse?'menu-bar-collapse-width':'menu-bar-width'">
    <img v-if="collapse" src="@/assets/logo.png"/> <div>{{collapse?'':appName}}</div>
  </div>
  <!-- 导航菜单 -->
  <el-menu ref="navmenu" default-active="1" :class="collapse?'menu-bar-collapse-width':"
    :collapse="collapse" :collapse-transition="false" :unique-opened="true"
    @open="handleopen" @close="handleclose" @select="handleselect">
    <!-- 导航菜单树组件，动态加载菜单 -->
    <menu-tree v-for="item in navTree" :key="item.id" :menu="item"></menu-tree>
  </el-menu>
</div>
```

图 28-5

# 28.5 页面测试

启动应用，访问 http://localhost:8080/#/login，登录进入主页，导航菜单已经显示出来，如图 28-6 所示。

图 28-6

# 第 29 章 页面权限控制

## 29.1 权限控制方案

既然是后台权限管理系统，当然少不了权限控制。至于权限控制，前端方面当然就是指对页面资源的访问和操作控制。前端资源权限主要分为两个部分，即导航菜单的查看权限和页面增、删、改操作按钮的操作权限。

我们的设计把页面导航菜单和页面操作按钮统一存储在菜单数据库表中，菜单表中包含以下权限关注点。

### 29.1.1 菜单类型

菜单类型代表页面资源的类型。

### 29.1.2 权限标识

权限标识是对页面资源进行权限控制的唯一标识，主要是增、删、改、查的权限控制。权限标识主要包含四种，以用户管理为例，权限标识包括 sys:user:add（新增）、sys:user:edit（编辑）、sys:user:delete（删除）、sys:user:view（查看）。

后台接口根据用户权限加载用户菜单数据返回给前端，前端导航菜单显示用户菜单数据，并在管理页面根据用户操作权限标识设置页面操作按钮的可见性或可用状态。我们这里采用无操作权限则页面操作按钮为不可用状态的方式。

### 29.1.3 菜单表结构

具体的菜单表结构如下，其中 type 是菜单类型，url 是菜单 URL，perms 是用户权限标识。

```
-- ----------------------------
-- Table structure for sys_menu
-- ----------------------------
DROP TABLE IF EXISTS `sys_menu`;
CREATE TABLE `sys_menu` (
```

```sql
`id` bigint(20) NOT NULL AUTO_INCREMENT COMMENT '编号',
`name` varchar(50) DEFAULT NULL COMMENT '菜单名称',
`parent_id` bigint(20) DEFAULT NULL COMMENT '父菜单ID,一级菜单为0',
`url` varchar(200) DEFAULT NULL COMMENT '菜单URL,类型：1.普通页面（如用户管理,/sys/user）2.嵌套完整外部页面,以http(s)开头的链接 3.嵌套服务器页面,使用iframe:前缀+目标URL(如SQL监控, iframe:/druid/login.html, iframe:前缀会替换成服务器地址)',
`perms` varchar(500) DEFAULT NULL COMMENT '授权(多个用逗号分隔,如：sys:user:add,sys:user:edit)',
`type` int(11) DEFAULT NULL COMMENT '类型   0：目录   1：菜单   2：按钮',
`icon` varchar(50) DEFAULT NULL COMMENT '菜单图标',
`order_num` int(11) DEFAULT NULL COMMENT '排序',
`create_by` varchar(50) DEFAULT NULL COMMENT '创建人',
`create_time` datetime DEFAULT NULL COMMENT '创建时间',
`last_update_by` varchar(50) DEFAULT NULL COMMENT '更新人',
`last_update_time` datetime DEFAULT NULL COMMENT '更新时间',
`del_flag` tinyint(4) DEFAULT '0' COMMENT '是否删除  -1：已删除   0：正常',
PRIMARY KEY (`id`)
) ENGINE=InnoDB AUTO_INCREMENT=45 DEFAULT CHARSET=utf8 COMMENT='菜单管理';
```

## 29.2 导航菜单实现思路

### 29.2.1 用户登录系统

用户登录成功之后跳转到首页。

### 29.2.2 根据用户加载导航菜单

在路由导航守卫路由时加载用户导航菜单并存储到 store。加载过程如下，返回结果排除按钮类型：user→user_role→role→role_menu→menu。

### 29.2.3 导航栏读取菜单树

导航栏页面到 store 读取导航菜单树并进行展示。

## 29.3 页面按钮实现思路

### 29.3.1 用户登录系统

用户登录成功之后跳转到首页。

### 29.3.2 加载权限标识

在路由导航守卫路由时加载用户权限标识集合并保存到 store 备用。加载过程如下，返回结果是用户权限标识的集合：user→user_role→role→role_menu→menu。

### 29.3.3 页面按钮控制

页面操作按钮提供 perms 属性绑定权限标识，使用 disable 属性绑定权限判断方法的返回值，权限判断方法 hasPerms(perms)通过查找上一步保存的用户权限标识集合是否包含 perms 来包含说明用户拥有此相关权限，否则设置当前操作按钮为不可用状态。

## 29.4 权限控制实现

### 29.4.1 导航菜单权限

#### 1. 加载导航菜单

在导航守卫路由时加载导航菜单并保存状态，如图 29-1 所示。

router/index.js

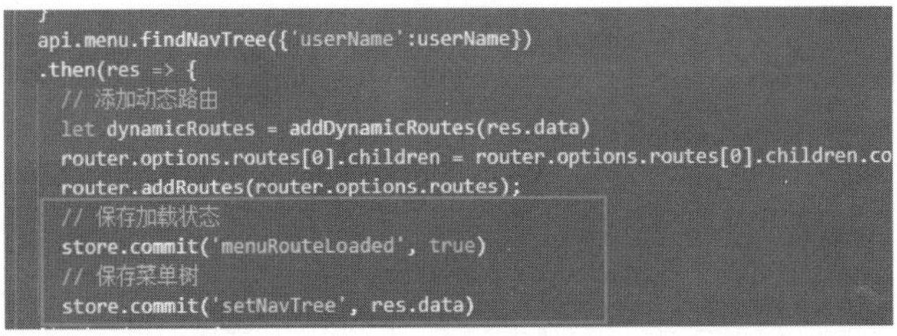

```
api.menu.findNavTree({'userName':userName})
.then(res => {
    // 添加动态路由
    let dynamicRoutes = addDynamicRoutes(res.data)
    router.options.routes[0].children = router.options.routes[0].children.co
    router.addRoutes(router.options.routes);
    // 保存加载状态
    store.commit('menuRouteLoaded', true)
    // 保存菜单树
    store.commit('setNavTree', res.data)
```

图 29-1

#### 2. 页面组件引用

导航栏页面从共享状态中读取导航菜单树并展示，如图 29-2 所示。

views/NavBar/NavBar.vue

```
computed: {
    ...mapState({
        appName: state=>state.app.appName,
        themeColor: state=>state.app.themeColor,
        collapse: state=>state.app.collapse,
        navTree: state=>state.menu.navTree
    })
}
```

图 29-2

修改 views/NavBar/NavBar.vue 文件，如图 29-3 所示。

```
<template>
  <div class="menu-bar-container">
    <!-- logo -->
    <div class="logo" :style="{'background-color':themeColor}" :class="co
        <img src="@/assets/logo.png"/> <div>{{collapse?'':appName}}</div>
    </div>
    <!-- 导航菜单 -->
    <el-menu default-active="1" :class="collapse?'menu-bar-collapse-width
        :collapse="collapse" @open="handleopen" @close="handleclose" @sele
        <!-- 导航菜单树组件，动态加载菜单 -->
        <menu-tree v-for="item in navTree" :key="item.id" :menu="item"></me
    </el-menu>
  </div>
</template>
```

图 29-3

## 29.4.2 页面按钮权限

### 1. 加载权限标识

打开 router/index.js，在导航守卫路由时加载权限标识并保存状态，如图 29-4 所示。

```
.then(res => {
    // 添加动态路由
    let dynamicRoutes = addDynamicRoutes(res.data)
    router.options.routes[0].children = router.options.routes[0].children.concat(
    router.addRoutes(router.options.routes);
    // 保存加载状态
    store.commit('menuRouteLoaded', true)
    // 保存菜单树
    store.commit('setNavTree', res.data)
}).then(res => {
    api.user.findPermissions({'name':userName}).then(res => {
        // 保存用户权限标识集合
        store.commit('setPerms', res.data)
    })
})
```

图 29-4

### 2. 权限按钮判断

封装了权限操作按钮组件，在组件中根据外部传入的权限标识进行权限判断。

```
views/Core/KtButton.vue
<template>
  <el-button :size="size" :type="type" :icon="icon" :loading="loading"
:disabled="!hasPerms(perms)" @click="handleClick"> {{label}}
  </el-button>
</template>

<script>
```

```js
import { hasPermission } from '@/permission/index.js'
export default {
  name: 'KtButton',
  props: {
    label: {  // 按钮显示文本
      type: String, default: 'Button'
    },
    icon: {  // 按钮显示图标
      type: String, default: ''
    },
    size: {  // 按钮尺寸
      type: String, default: 'mini'
    },
    type: {  // 按钮类型
      type: String, default: null
    },
    loading: {  // 按钮加载标识
      type: Boolean, default: false
    },
    disabled: {  // 按钮是否禁用
      type: Boolean, default: false
    },
    perms: {  // 按钮权限标识,外部使用者传入
      type: String, default: null
    }
  },
  methods: {
    handleClick: function () {
      // 按钮操作处理函数
      this.$emit('click', {})
    },
    hasPerms: function (perms) {
      // 根据权限标识和外部指示状态进行权限判断
      return hasPermission(perms) & !this.disabled
    }
  }
}
</script>
```

### 3. 权限判断逻辑

新建权限文件 src/permission/index.js,提供权限判断方法 hasPermission,传入当前组件绑定的权限标识 perms,判断权限标识是否存在于 store 中保存的用户权限标识集合中。

```
src/permission/index.js
import store from '@/store'
/**
 * 判断用户是否拥有操作权限
 * 根据传入的权限标识,查看是否存在用户权限标识集合
 * @param perms
 */
export function hasPermission (perms) {
    let hasPermission = false
    let permissions = store.state.user.perms
    for(let i=0, len=permissions.length; i<len; i++) {
        if(permissions[i] === perms) {
            hasPermission = true;
            break
        }
    }
    return hasPermission
}
```

### 4. 权限按钮引用

在文件 views/Sys/User.vue 中配置相关内容,新建一个用户管理页面。在页面中引入权限按钮组件并绑定权限标识,如图 29-5 所示。

views/Sys/User.vue

```
<!--工具栏-->
<div class="toolbar" style="float:left; padding:18px;">
    <el-form :inline="true" :model="filters" size="small">
        <el-form-item>
            <el-input v-model="filters.name" placeholder="用户名"></el-input>
        </el-form-item>
        <el-form-item>
            <el-button type="primary" v-on:click="findPage(null)">查询</el-button>
        </el-form-item>
        <el-form-item>
            <kt-button label="新增" perms="sys:user:add" type="primary" @click="handleAdd" />      外部使用
        </el-form-item>
    </el-form>
</div>                        表格进行了二次封装,所以得先把权限标识传入表格,再由表格传入权限按钮
<!--表格内容栏-->
<kt-table permsEdit="sys:user:edit" permsDelete="sys:user:delete"
    :data="pageResult" :columns="columns"
    @findPage="findPage" @handleEdit="handleEdit" @handleDelete="handleDelete">
</kt-table>
```

图 29-5

在文件 views/Core/ KtTable.vue 中配置相关内容,添加一个表格封装组件,在组件中引入权限按钮组件,并在表格操作涉及的编辑、删除、批量删除按钮绑定对应的权限标识,如图 29-6 所示。

views/Core/ KtTable.vue

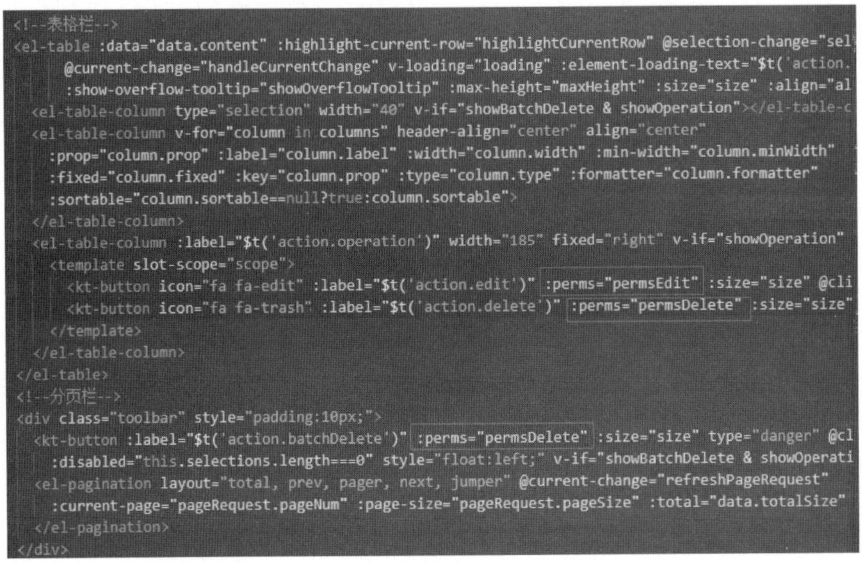

```html
<!--表格栏-->
<el-table :data="data.content" :highlight-current-row="highlightCurrentRow" @selection-change="sel
    @current-change="handleCurrentChange" v-loading="loading" :element-loading-text="$t('action.
    :show-overflow-tooltip="showOverflowTooltip" :max-height="maxHeight" :size="size" :align="al
    <el-table-column type="selection" width="40" v-if="showBatchDelete & showOperation"></el-table-c
    <el-table-column v-for="column in columns" header-align="center" align="center"
        :prop="column.prop" :label="column.label" :width="column.width" :min-width="column.minWidth"
        :fixed="column.fixed" :key="column.prop" :type="column.type" :formatter="column.formatter"
        :sortable="column.sortable==null?true:column.sortable">
    </el-table-column>
    <el-table-column :label="$t('action.operation')" width="185" fixed="right" v-if="showOperation"
        <template slot-scope="scope">
            <kt-button icon="fa fa-edit" :label="$t('action.edit')" :perms="permsEdit" :size="size" @cli
            <kt-button icon="fa fa-trash" :label="$t('action.delete')" :perms="permsDelete" :size="size
        </template>
    </el-table-column>
</el-table>
<!--分页栏-->
<div class="toolbar" style="padding:10px;">
    <kt-button :label="$t('action.batchDelete')" :perms="permsDelete" :size="size" type="danger" @cl
        :disabled="this.selections.length===0" style="float:left;" v-if="showBatchDelete & showOperati
    <el-pagination layout="total, prev, pager, next, jumper" @current-change="refreshPageRequest"
        :current-page="pageRequest.pageNum" :page-size="pageRequest.pageSize" :total="data.totalSize"
    </el-pagination>
</div>
```

图 29-6

## 29.5 标签页功能

新建 store/modules/tab.js 文件，用于存储标签页和当前选中标签选项。

```
store/modules/tab.js
export default {
  state: {
    // 主入口标签页
    mainTabs: [],
    // 当前标签页名
    mainTabsActiveName: ''
  },
  mutations: {
    updateMainTabs (state, tabs) {
      state.mainTabs = tabs
    },
    updateMainTabsActiveName (state, name) {
      state.mainTabsActiveName = name
    }
  }
}
```

修改 views/MainContent.vue 文件，在主内容区域上面添加标签页组件，通过 Element 的选项卡 el-tabs 组件实现，能够通过切换标签页查看不同路由页面，并需要支持导航菜单和标签页同步，如图 29-7 所示。

```
<div id="main-container" class="main-container" :class="$store.state.app.collapse?'position-collapse
  <!-- 标签页 -->
  <div class="tab-container">
    <el-tabs class="tabs" :class="$store.state.app.collapse?'position-collapse-left':'position-left
      v-model="mainTabsActiveName" :closable="true" type="card"
      @tab-click="selectedTabHandle" @tab-remove="removeTabHandle">
      <el-dropdown class="tabs-tools" :show-timeout="0" trigger="hover">
        <div style="font-size:20px;width:50px;"><i class="el-icon-arrow-down"></i></div>
        <el-dropdown-menu slot="dropdown">
          <el-dropdown-item @click.native="tabsCloseCurrentHandle">关闭当前标签</el-dropdown-item>
          <el-dropdown-item @click.native="tabsCloseOtherHandle">关闭其它标签</el-dropdown-item>
          <el-dropdown-item @click.native="tabsCloseAllHandle">关闭全部标签</el-dropdown-item>
          <el-dropdown-item @click.native="tabsRefreshCurrentHandle">刷新当前标签</el-dropdown-item>
        </el-dropdown-menu>
      </el-dropdown>
      <el-tab-pane v-for="item in mainTabs"
        :key="item.name" :label="item.title" :name="item.name">
        <span slot="label"><i :class="item.icon"></i> {{item.title}} </span>
      </el-tab-pane>
    </el-tabs>
  </div>
  <!-- 主内容区域 -->
  <div class="main-content">
```

图 29-7

通过 watch 实现对路由对象$route 的监听，监听到路由变化的时候调用 handleRoute 方法进行处理，处理逻辑主要包括 3 项：① 判断标签页是否存在，不存在则创建；② 如果是新创建了标签页，就把新标签页加入当前标签页集合；③ 同步标签页路由到导航菜单，设置选中状态。

```
watch: {
  $route: 'handleRoute'
},
methods: {
  // 路由操作处理
  handleRoute (route) {
    // tab 标签页选中，如果不存在就先添加
    var tab = this.mainTabs.filter(item => item.name === route.name)[0]
    if (!tab) {
      tab = {
        name: route.name,
        title: route.name,
        icon: route.meta.icon
      }
      this.mainTabs = this.mainTabs.concat(tab)
    }
    this.mainTabsActiveName = tab.name
    // 切换标签页时同步更新高亮菜单
    if(this.$refs.navmenu != null) {
```

```
      this.$refs.navmenu.activeIndex = '' + route.meta.index
      this.$refs.navmenu.initOpenedMenu()
    }
  }
 }
}
```

上一步将标签页存储到了 store，这里 mainTabs 从 store 中获取标签页数据提供给页面。

```
computed: {
  mainTabs: {
    get () { return this.$store.state.tab.mainTabs },
    set (val) { this.$store.commit('updateMainTabs', val) }
  },
  mainTabsActiveName: {
    get () { return this.$store.state.tab.mainTabsActiveName },
    set (val) { this.$store.commit('updateMainTabsActiveName', val) }
  }
}
```

标签页对应的操作函数包括标签页的选中、关闭等系列操作。

```
  // tabs, 选中tab
  selectedTabHandle (tab) {
    tab = this.mainTabs.filter(item => item.name === tab.name)
    if (tab.length >= 1) {
      this.$router.push({ name: tab[0].name })
    }
  },
  // tabs, 删除tab
  removeTabHandle (tabName) {
    this.mainTabs = this.mainTabs.filter(item => item.name !== tabName)
    if (this.mainTabs.length >= 1) {
      // 当前选中tab被删除
      if (tabName === this.mainTabsActiveName) {
        this.$router.push({ name: this.mainTabs[this.mainTabs.length - 1].name },
() => {
          this.mainTabsActiveName = this.$route.name
        })
      }
    } else {
      this.$router.push("/")
    }
  },
  // tabs, 关闭当前
```

```js
    tabsCloseCurrentHandle () {
      this.removeTabHandle(this.mainTabsActiveName)
    },
    // tabs, 关闭其他
    tabsCloseOtherHandle () {
      this.mainTabs = this.mainTabs.filter(item => item.name === this.mainTabsActiveName)
    },
    // tabs, 关闭全部
    tabsCloseAllHandle () {
      this.mainTabs = []
      this.$router.push("/")
    },
    // tabs, 刷新当前
    tabsRefreshCurrentHandle () {
      var tempTabName = this.mainTabsActiveName
      this.removeTabHandle(tempTabName)
      this.$nextTick(() => {
        this.$router.push({ name: tempTabName })
      })
    }
```

## 29.6 系统介绍页

为方便标签页效果的测试,这里添加一个系统介绍页,作为系统的首页。

新建 views/Intro/Intro.vue 文件结构,添加系统介绍页面,附上介绍信息,如图 29-8 所示。

views/Intro/Intro.vue

```html
<template>
  <div class="page-container" style="width:99%;margin-top:15px;">
    <el-carousel :interval="3000" type="card" height="450px" class="carousel">
      <el-carousel-item class="carousel-item-intro">
        <h2>项目介绍</h2>
        <ul>
          <li>基于 Spring Boot、Spring Cloud、Vue、Element 的 Java EE 快速开发平台</li>
          <li>旨在提供一套简洁易用的解决方案,帮助用户有效降低项目开发难度和成本</li>
          <li>博客提供项目开发过程同步系列教程文章,手把手的教你如何开发同类系统</li>
        </ul>
        <div><img src="@/assets/logo.png" style="width:120px;height:120px;padding-top:15px;" /></div>
      </el-carousel-item>
      <el-carousel-item class="carousel-item-func">
        <h2>功能计划</h2>
        <ul>
          <li>✓ 系统登录:系统用户登录,系统登录认证(token方式)</li>
          <li>✓ 用户管理:新建用户,修改用户,删除用户,查询用户</li>
          <li>✓ 机构管理:新建机构,修改机构,删除机构,查询机构</li>
          <li>✓ 角色管理:新建角色,修改角色,删除角色,查询角色</li>
          <li>✓ 菜单管理:新建菜单,修改菜单,删除菜单,查询菜单</li>
```

图 29-8

在 router/index.js 中添加系统介绍页的路由。router/index.js 内容如图 29-9 所示。

图 29-9

## 29.7 页面测试

启动应用，访问 http://localhost:8080/#/login，登录后进入主页，默认显示系统介绍页，如图 29-10 所示。

图 29-10

打开导航菜单"系统管理"→"用户管理"，可以看到打开的标签页和用户管理界面，如图 29-11 所示。

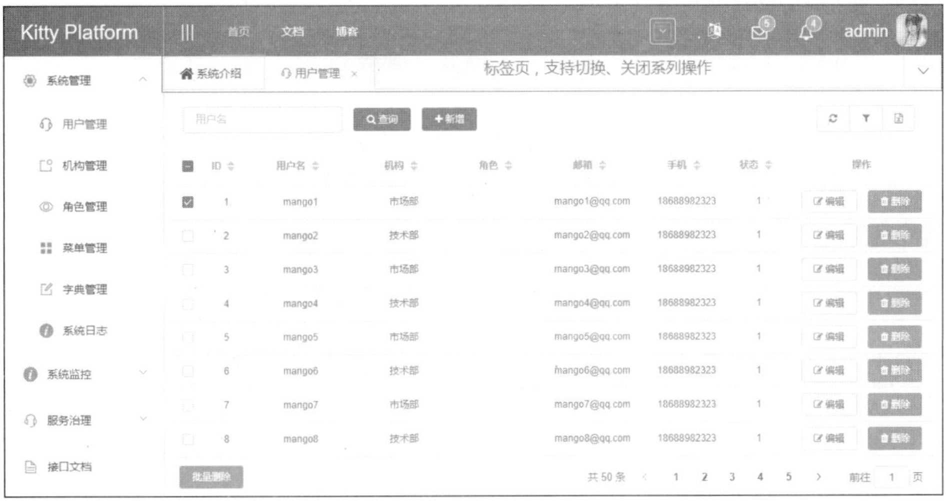


图 29-11

我们测试一下权限按钮，修改用户管理界面的新增按钮的权限标识，把它从有效的“sys:user:add”改为无效的“sys:user:add2”。

views/Sys/user.vue 内容如图 29-12 所示。

```
div class="page-container">
 <!--工具栏-->
 <div class="toolbar" style="float:left;padding-top:10px;padding-left:15px;">
     <el-form :inline="true" :model="filters" :size="size">
         <el-form-item>
             <el-input v-model="filters.name" placeholder="用户名"></el-input>
         </el-form-item>
         <el-form-item>
             <kt-button icon="fa fa-search" :label="$t('action.search')" perms="sys:role:view" type=
         </el-form-item>
         <el-form-item>
             <kt-button icon="fa fa-plus" :label="$t('action.add')" perms="sys:user:add2" type="prim
         </el-form-item>
     </el-form>
 </div>
```

图 29-12

退出系统重新登录，再次打开用户管理界面，发现因为用户缺少权限，“新增”按钮已变为不可用状态（见图 29-13），这样就实现了页面按钮的权限控制。

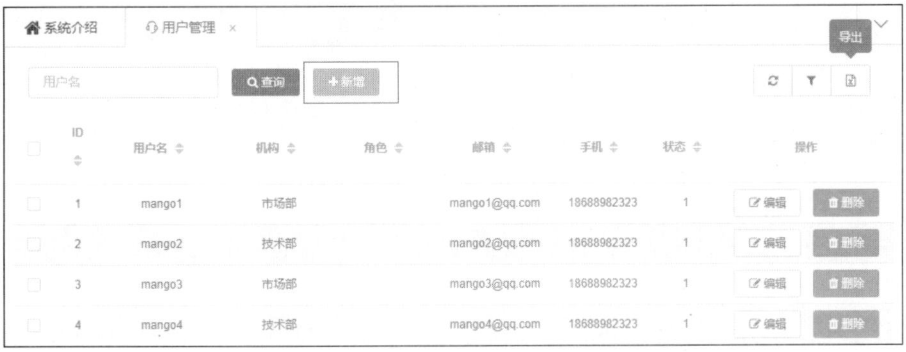


图 29-13

# 第 30 章 功能管理模块

就目前来看，功能管理页面大多类似，如用户管理、功能管理模块中的字典管理、系统配置、登录日志和操作日志等都是以表格管理数据为主，机构管理和菜单管理则以表格树的数据管理为主，所以这里在每个类型中挑选一个作为讲解案例，其他页面不再复述，读者用到的时候查阅相关代码即可。

## 30.1 字典管理

上一章我们已经实现了用户管理模块，功能管理模块中的字典管理、系统配置、登录日志、和操作日志等功能模块都是类似用户管理模块的功能，以表格展示数据为主，且登录日志和操作日志只提供可读、不提供编辑和删除功能。

接下来，我们以字典管理为例再次讲解一下表格功能模块的实现，其他模块就不再赘述了。

### 30.1.1 关键代码

参照用户管理界面，在 views/Sys 下添加字段管理页面组件，逐步添加页面内容。

（1）添加工具栏内容。

```
views/Sys/Dict.vue
    <!--工具栏-->
    <div class="toolbar" style="float:left;padding-top:10px;padding-left:15px;">
        <el-form :inline="true" :model="filters" :size="size">
            <el-form-item>
                <el-input v-model="filters.label" placeholder="名称"></el-input>
            </el-form-item>
            <el-form-item>
                <kt-button icon="fa fa-search" :label="$t('action.search')" perms="sys:dict:view" type="primary" @click="findPage(null)"/>
            </el-form-item>
            <el-form-item>
                <kt-button icon="fa fa-plus" :label="$t('action.add')" perms="sys:dict:add" type="primary" @click="handleAdd" />
```

```
        </el-form-item>
     </el-form>
  </div>
```

（2）添加展示表格 kt-table 组件。

views/Sys/Dict.vue
```
     <!--表格内容栏-->
     <kt-table permsEdit="sys:dict:edit" permsDelete="sys:dict:delete"
         :data="pageResult" :columns="columns"
         @findPage="findPage" @handleEdit="handleEdit" @handleDelete="handleDelete">
     </kt-table>
```

（3）编辑对话框。

views/Sys/Dict.vue
```
     <!--新增编辑界面-->
     <el-dialog :title="operation?'新增':'编辑'"
width="40%" :visible.sync="editDialogVisible" :close-on-click-modal="false">
         <el-form :model="dataForm" label-width="80px" :rules="dataFormRules" ref="dataForm" :size="size">
             <el-form-item label="ID" prop="id" v-if="false">
                 <el-input v-model="dataForm.id" :disabled="true" auto-complete="off"></el-input>
             </el-form-item>
             <el-form-item label="名称" prop="label">
                 <el-input v-model="dataForm.label" auto-complete="off"></el-input>
             </el-form-item>
             <el-form-item label="值" prop="value">
                 <el-input v-model="dataForm.value" auto-complete="off"></el-input>
             </el-form-item>
             <el-form-item label="类型" prop="type">
                 <el-input v-model="dataForm.type" auto-complete="off"></el-input>
             </el-form-item>
             <el-form-item label="排序" prop="sort">
                 <el-input v-model="dataForm.sort" auto-complete="off"></el-input>
             </el-form-item>
             <el-form-item label="描述 " prop="description">
                 <el-input v-model="dataForm.description" auto-complete="off" type="textarea"></el-input>
             </el-form-item>
             <el-form-item label="备注" prop="remarks">
```

```
            <el-input v-model="dataForm.remarks" auto-complete="off"
type="textarea"></el-input>
          </el-form-item>
      </el-form>
      <div slot="footer" class="dialog-footer">
          <el-button :size="size" @click.native="editDialogVisible =
false">{{$t('action.cancel')}}</el-button>
          <el-button :size="size" type="primary"
@click.native="submitForm" :loading="editLoading">{{$t('action.submit')}}</el-button>
      </div>
    </el-dialog>
```

（4）编写分页查询方法。

```
views/Sys/Dict.vue
// 获取分页数据
findPage: function (data) {
if(data !== null) {
      this.pageRequest = data.pageRequest
    }
    this.pageRequest.params = [{name:'label', value:this.filters.label}]
    this.$api.dict.findPage(this.pageRequest).then((res) => {
      this.pageResult = res.data
    }).then(data!=null?data.callback:'')
}
```

（5）编写批量删除方法。

```
views/Sys/Dict.vue
// 批量删除
handleDelete: function (data) {
this.$api.dict.batchDelete(data.params).then(data!=null?data.callback:'')
}
```

（6）编辑提交方法。

```
views/Sys/Dict.vue
submitForm: function () {
    this.$refs.dataForm.validate((valid) => {
        if (valid) {
            this.$confirm('确认提交吗？', '提示', {}).then(() => {
                this.editLoading = true
                let params = Object.assign({}, this.dataForm)
                this.$api.dict.save(params).then((res) => {
                    if(res.code == 200) {
                        this.$message({ message: '操作成功',
                            type: 'success' })
```

```
            } else {
                this.$message({message: '操作失败, ' +
                    res.msg, type: 'error'})
            }
            this.editLoading = false
            this.$refs['dataForm'].resetFields()
            this.editDialogVisible = false
            this.findPage(null)
          })
        })
      }
    })
}
```

### 30.1.2　页面截图

字典管理页面如图 30-1 所示。

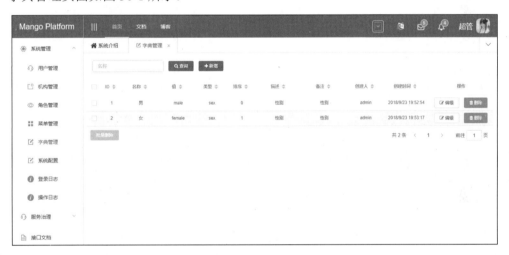

图 30-1

# 30.2　角色管理

角色管理页面除了表格展示之外，还可以给角色赋予角色菜单。接下来，我们就讲解一下角色管理页面的实现。

### 30.2.1　关键代码

首先在 views/Sys 下新建角色管理页面，然后模仿用户管理加入表格组件，再在角色表格下面添加一个角色菜单树，为用户分配角色菜单。角色菜单树组件如下：

views/Sys/Role.vue
```
    <!--角色菜单，表格树内容栏-->
    <div class="menu-container" :v-if="true">
        <div class="menu-header">
            <span><B>角色菜单授权</B></span>
        </div>
        <el-tree :data="menuData" size="mini" show-checkbox
node-key="id" :props="defaultProps"
            style="width: 100%;pading-top:20px;"
ref="menuTree" :render-content="renderContent"
            v-loading="menuLoading" element-loading-text="拼命加载中
" :check-strictly="true"
            @check-change="handleMenuCheckChange">
        </el-tree>
        <div
style="float:left;padding-left:24px;padding-top:12px;padding-bottom:4px;">
            <el-checkbox v-model="checkAll"
@change="handleCheckAll" :disabled="this.selectRole.id == null"><b>全选
</b></el-checkbox>
        </div>
        <div
style="float:right;padding-right:15px;padding-top:4px;padding-bottom:4px;">
            <kt-button :label="$t('action.reset')" perms="sys:role:edit"
type="primary" @click="resetSelection"
                :disabled="this.selectRole.id == null"/>
            <kt-button :label="$t('action.submit')" perms="sys:role:edit"
type="primary" @click="submitAuthForm"
                :disabled="this.selectRole.id == null" :loading="authLoading"/>
        </div>
    </div>
```

然后针对选中的角色同步更新菜单树的勾选项以及进行菜单树本身节点选中时对应父子关系的同步选中状态变更。

views/Sys/Role.vue
```
// 角色选择改变监听
    handleRoleSelectChange(val) {
        if(val == null || val.val == null) {
            return
        }
        this.selectRole = val.val
        this.$api.role.findRoleMenus({'roleId':val.val.id}).then((res) => {
            this.currentRoleMenus = res.data
            this.$refs.menuTree.setCheckedNodes(res.data)
        })
    },
```

```js
        // 树节点选择监听
        handleMenuCheckChange(data, check, subCheck) {
            if(check) {
                // 节点选中时同步选中父节点
                let parentId = data.parentId
                this.$refs.menuTree.setChecked(parentId, true, false)
            } else {
                // 节点取消选中时同步取消选中子节点
                if(data.children != null) {
                    data.children.forEach(element => {
                        this.$refs.menuTree.setChecked(element.id, false, false)
                    });
                }
            }
        }
```

## 30.2.2 页面截图

角色管理页面如图 30-2 所示。

图 30-2

# 30.3 菜单管理

菜单管理页面主要是放置一个表格树组件，可以支持表格树的新增、编辑和删除。机构管理页面与此类似，所以这里以菜单管理为例进行讲解。

在 Core 目录下添加一个 TableTreeColumn 封装组件，作为表格中可展开的列。

## 30.3.1 表格列组件

```
views/Core/TableTreeColumn.vue
<template>
  <el-table-column :prop="prop" v-bind="$attrs">
    <template slot-scope="scope">
      <span @click.prevent="toggleHandle(scope.$index, scope.row)" :style="childStyles(scope.row)">
        <i :class="iconClasses(scope.row)" :style="iconStyles(scope.row)"></i>
        {{ scope.row[prop] }}
      </span>
    </template>
  </el-table-column>
</template>

<script>
  import isArray from 'lodash/isArray'
  export default {
    name: 'table-tree-column',
    props: { // 省略 props },
    methods: {
      childStyles (row) {
        return { 'padding-left': (row[this.levelKey] * 25) + 'px' }
      },
      iconClasses (row) {
        return [ !row._expanded ? 'el-icon-caret-right':'el-icon-caret-bottom' ]
      },
      iconStyles (row) {
        return { 'visibility': this.hasChild(row) ? 'visible' : 'hidden' }
      },
      hasChild (row) {
        return(isArray(row[this.childKey])&&row[this.childKey].length>=1)||false
      },
      // 切换处理
      toggleHandle (index, row) {
        if (this.hasChild(row)) {
          var data = this.$parent.store.states.data.slice(0)
          data[index]._expanded = !data[index]._expanded
          if (data[index]._expanded) {
            data=data.splice(0, index+1).concat(row[this.childKey]).concat(data)
          } else {
```

```
        data = this.removeChildNode(data, row[this.treeKey])
      }
      this.$parent.store.commit('setData', data)
      this.$nextTick(() => { this.$parent.doLayout()})
    }
  },
  // 移除子节点
  removeChildNode (data, parentId) {
    var parentIds = isArray(parentId) ? parentId : [parentId]
    if (parentId.length <= 0) { return data }
    var ids = []
    for (var i = 0; i < data.length; i++) {
      if (parentIds.indexOf(data[i][this.parentKey]) !== -1
&& parentIds.indexOf(data[i][this.treeKey]) === -1) {
        ids.push(data.splice(i, 1)[0][this.treeKey])
        i--
      }
    }
    return this.removeChildNode(data, ids)
  }
}
}
</script>
```

## 30.3.2 创建表格树

在 Sys 目录下新建菜单管理页面，加入工具栏、编辑对话框与其他功能类型，主要是在页面放置表格树组件，放置一个 el-table，针对 name 列使用我们封装的 TableTreeColumn 组件显示子树的展示。

```
views/Sys/Menu.vue
    <!--表格树内容栏-->
    <el-table :data="tableTreeDdata" stripe size="mini" style="width: 100%;"
      v-loading="loading" element-loading-text="$t('action.loading')">
      <el-table-column
        prop="id" header-align="center" align="center" width="80" label="ID">
      </el-table-column>
      <table-tree-column
        prop="name" header-align="center" treeKey="id" width="150" label="名称">
      </table-tree-column>
      <el-table-column header-align="center" align="center" label="图标">
        <template slot-scope="scope">
          <i :class="scope.row.icon || ''"></i>
        </template>
```

```
        </el-table-column>
        // 省略其他列声明
</el-table>
```

### 30.3.3 页面截图

菜单管理页面如图 30-3 所示。

图 30-3

# 第 31 章 嵌套外部网页

## 31.1 需求背景

有些时候，我们需要内嵌外部网页，可以通过单击导航菜单，然后将我们的主内容栏加载外部网页的内容进行显示，如查看服务端提供的 SQL 监控页面、接口文档页面等。

这时就要求我们的导航菜单能够解析外部嵌套网页的 URL，并根据 URL 路由到相应的嵌套组件。接下来我们讲解具体实现方案。

## 31.2 实现原理

（1）给菜单 URL 添加外部嵌套网页格式，除内部渲染的页面 URL 外，外部网页统一直接以 http[s]完整路径开头。

（2）路由导航守卫在动态加载路由时，检测到如果是外部嵌套网页，就绑定到 IFrame 嵌套组件，最后使用 IFrame 组件来渲染嵌套页面。

（3）菜单单击跳转的时候，根据路由类型生成不同的路由路径，载入特定的页面内容渲染到步骤（2）绑定的特定组件上。

## 31.3 代码实现

前面的原理听起来似乎有点笼统，我们来看看具体的实现过程。

### 31.3.1 确定菜单 URL

服务监控页面，其实显示的就是服务监控提供的现有页面。访问地址是 http://localhost:8001/druid/login.html，即完整 HTTP 地址格式。效果如图 31-1 所示，输入服务端配置的账号、密码就可以查看了。

账号信息是“用户名：admin”“密码：admin”。

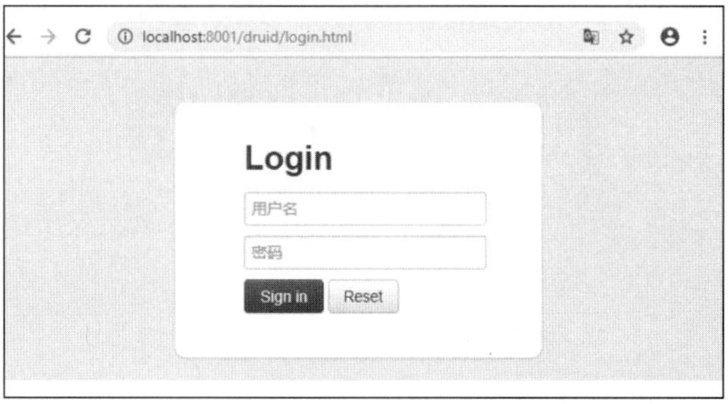


图 31-1

登录之后，可以看到各种数据库相关的监控记录，是数据库监控和调优的利器，如图 31-2 所示。

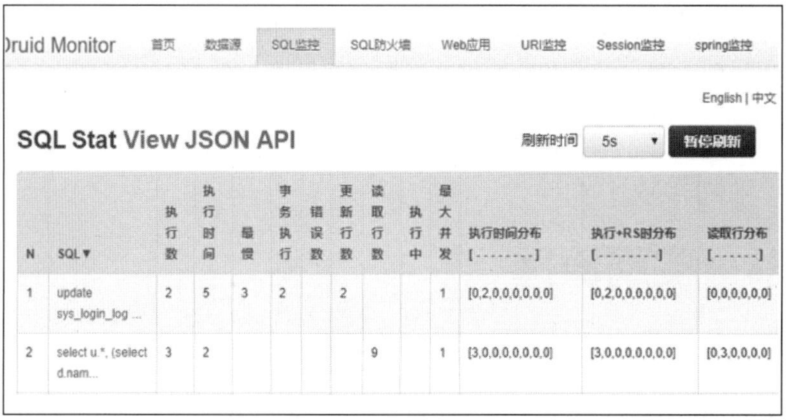


图 31-2

我们将 SQL 监控的菜单 URL 配置为 http://127.0.0.1:8001/druid/login.html，localhost 跟 127.0.0.1 是一个意思，代表本机，届时路由解析时检测到以 http 开头的就是外部网页，然后绑定到 IFrame 嵌套页面组件上进行渲染，如图 31-3 所示。

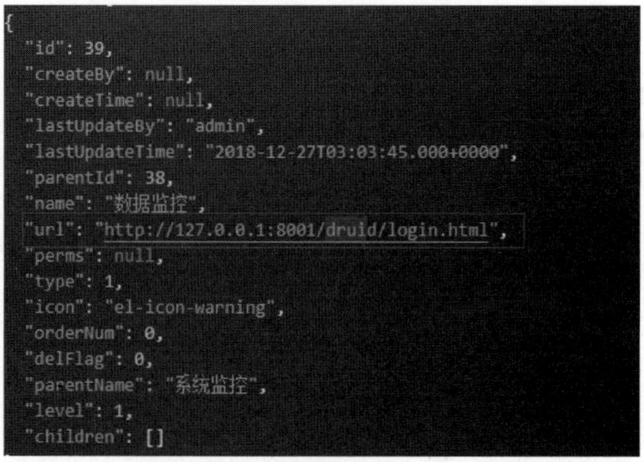

```
{
  "id": 39,
  "createBy": null,
  "createTime": null,
  "lastUpdateBy": "admin",
  "lastUpdateTime": "2018-12-27T03:03:45.000+0000",
  "parentId": 38,
  "name": "数据监控",
  "url": "http://127.0.0.1:8001/druid/login.html",
  "perms": null,
  "type": 1,
  "icon": "el-icon-warning",
  "orderNum": 0,
  "delFlag": 0,
  "parentName": "系统监控",
  "level": 1,
  "children": []
```


图 31-3

### 31.3.2 创建嵌套组件

在 views 下创建 IFrame 目录并在其下新建一个 IFrame.vue 嵌套组件。

IFrame 组件在渲染时，读取 store 的 iframeUrl 以加载要渲染的内容（通过设置 src）。

```
views/IFrame/IFrame.vue
<template>
  <div class="iframe-container">
<iframe :src="src" scrolling="auto" frameborder="0" class="frame" :onload="onloaded()">
   </iframe>
  </div>
</template>

<script>
export default {
  data() {
    return { src: "", loading: null }
  },
  methods: {
    resetSrc: function(url) {
      this.src = url
      this.load()
    },
    load: function() {
      this.loading = this.$loading({
        lock: true, text: "loading...", spinner: "el-icon-loading",
        background: "rgba(0, 0, 0, 0.5)",
        target: document.querySelector("#main-container ")
      })
    },
    onloaded: function() {
      if(this.loading) { this.loading.close() }
    }
  },
  mounted() {
    this.resetSrc(this.$store.state.iframe.iframeUrl);
  },
  watch: {
    $route: {
      handler: function(val, oldVal) {
        // 如果是跳转到嵌套页面，切换 iframe 的 url
        this.resetSrc(this.$store.state.iframe.iframeUrl);
```

```
        }
      }
    }
  }
}
</script>

// 此处省略 CSS
```

### 31.3.3 绑定嵌套组件

在导航守卫动态加载路由的时候，解析 URL，如果是嵌套页面，就绑定到 IFrame 组件。router/index.js 的相关内容如图 31-4 所示。

```
// 创建路由配置
var route = {
  path: menuList[i].url,
  component: null,
  name: menuList[i].name,
  meta: {
    icon: menuList[i].icon,
    index: menuList[i].id
  }
}
let path = getIFramePath(menuList[i].url)
if (path) {
  // 如果是嵌套页面，通过iframe展示
  route['path'] = path
  route['component'] = resolve => require([`@/views/IFrame/IFrame`], resolve)
  // 存储嵌套页面路由路径和访问URL
  let url = getIFrameUrl(menuList[i].url)
  let iframeUrl = {'path':path, 'url':url}
  store.commit('addIFrameUrl', iframeUrl)
} else {
```

图 31-4

在每次路由时，把路由路径保存到 store，如果是 IFrame 嵌套页面，IFrame 就会在渲染时到 store 中读取 iframeUrl 以确定渲染的内容。

```
router/index.js
/**
 * 处理 IFrame 嵌套页面
 */
function handleIFrameUrl(path) {
  // 嵌套页面，保存 iframeUrl 到 store，供 IFrame 组件读取展示
  let url = path
  let length = store.state.iframe.iframeUrls.length
  for(let i=0; i<length; i++) {
    let iframe = store.state.iframe.iframeUrls[i]
    if(path != null && path.endsWith(iframe.path)) {
      url = iframe.url
      store.commit('setIFrameUrl', url)
```

```
            break
        }
    }
}
```

在 sotre/modules 下新建 iframe.js 文件,存储 IFrame 状态。

```
sotre/modules/iframe.js
export default {
    state: {
        iframeUrl: [],      // 当前嵌套页面路由路径
        iframeUrls: []      // 所有嵌套页面路由路径访问 URL
    },
    getters: {
    },
    mutations: {
        setIFrameUrl(state, iframeUrl){  // 设置 iframeUrl
            state.iframeUrl = iframeUrl
        },
        addIFrameUrl(state, iframeUrl){  // iframeUrls
            state.iframeUrls.push(iframeUrl)
        }
    },
    actions: {
    }
}
```

iframe.js 是一个工具类,主要对嵌套 URL 进行处理。

```
utils/iframe.js
/**
 * 嵌套页面 IFrame 模块
 */

import { baseUrl } from '@/utils/global'

/**
 * 嵌套页面 URL 地址
 * @param {*} url
 */
export function getIFramePath (url) {
    let iframeUrl = ''
    if(/^iframe:.*/.test(url)) {
        iframeUrl = url.replace('iframe:', '')
    } else if(/^http[s]?:\/\/.*/.test(url)) {
```

```
      iframeUrl = url.replace('http://', '')
      if(iframeUrl.indexOf(":") != -1) {
        iframeUrl = iframeUrl.substring(iframeUrl.lastIndexOf(":") + 1)
      }
    }
    return iframeUrl
}

/**
 * 嵌套页面路由路径
 * @param {*} url
 */
export function getIFrameUrl (url) {
  let iframeUrl = ''
  if(/^iframe:.*/.test(url)) {
    iframeUrl = baseUrl + url.replace('iframe:', '')
  } else if(/^http[s]?:\/\/.*/.test(url)) {
    iframeUrl = url
  }
  return iframeUrl
}
```

### 31.3.4　菜单路由跳转

在菜单路由跳转的时候，判断是否是 IFrame 路由，如果是就处理成 IFrame 需要的路由 URL 进行跳转。

```
components/MenuTree/index.vue
handleRoute (menu) {
  // 如果是嵌套页面, 转换成 iframe 的 path
  let path = getIFramePath(menu.url)
  if(!path) {
    path = menu.url
  }
  // 通过菜单 URL 跳转至指定路由
  this.$router.push("/" + path)
}
```

## 31.4　页面测试

启动注册中心、监控服务、后台服务，访问 http://localhost:8080，登录主页。

单击导航栏"系统监控"→"数据监控",可以看到在主内容区域加载了数据监控页面内容。

 如果页面显示拒绝加载,查看控制台是否是因为 X-Frame-Options 的设置。因为 Spring Security 默认是不允许页面被嵌套的,所以 X-Frame-Options 被默认设置为 DENY,这个可以在后台 WebSecurityConfig 配置类通过 http.headers().frameOptions().disable()。

禁用 X-Frame-Options 设置就可以正常显示了,如图 31-5 所示。

图 31-5

同理,实现其他嵌套页面,服务监控页面如图 31-6 所示。

图 31-6

接口文档页面如图 31-7 所示。

图 31-7

注册中心页面如图 31-8 所示。

图 31-8

# 第 32 章 数据备份还原

## 32.1 需求背景

在很多时候，我们需要对系统数据进行备份还原。当然，生产线上专业的备份还原会由专门的工作人员如 DBA 在数据库服务端直接进行备份还原。我们这里主要是通过界面进行少量的数据备份和还原，比如做一个在线演示系统，为了方便恢复被演示用户删除或修改过的演示数据，我们这里就实现了这么一个系统数据备份和还原的功能，正好可以给大家作为示例进行讲解。我们通过代码调用 MySQL 的备份还原命令实现系统备份还原的功能，具体逻辑都是由后台代码实现的，参见数据备份还原后端实现篇。

## 32.2 后台接口

在前期我们已经准备好了系统备份还原的后台接口，具体可以参考备份还原后台篇。

备份还原主要接口有以下几个。

- backup：备份创建接口，会在服务端_backup 目录下生成以时间戳相关的备份目录，目录下有 MySQL 的备份 SQL。
- delete：系统备份删除接口，传入页面查询得到的备份名称作为参数，删除服务端备份记录。
- findRecord：系统备份查询接口，查询所有备份记录，返回给前台页面展示，用于还原和删除。
- restore：系统备份还原接口，传入页面查询得到的备份名称作为参数，还原系统数据到当前备份。

可以访问备份服务的 Swagger 查看接口文档，如图 32-1 所示。

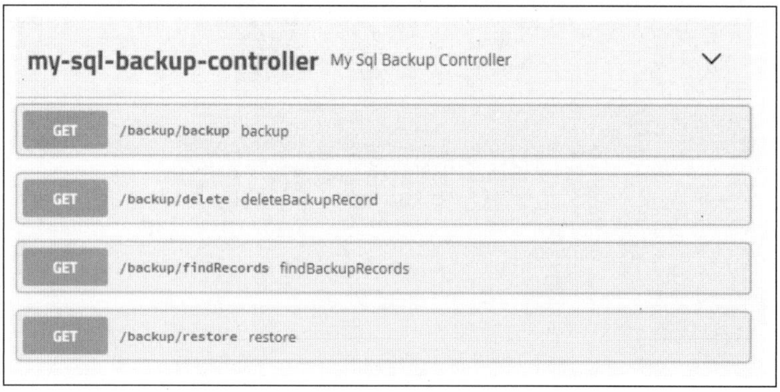

图 32-1

## 32.3 备份页面

在 views 目录下新建 Backup 目录和页面。页面内容主要是一个可以进行备份和还原的对话框组件。

### 1. 模板内容

```
views/Backup/Backup.vue
<template>
    <!--备份还原界面-->
    <el-dialog :title="$t('common.backupRestore')" width="40%" :visible.sync="backupVisible" :close-on-click-modal="false" :modal=false>
        <el-table :data="tableData" style="width: 100%;font-size:16px;" height="330px" :show-header="showHeader"
            size="mini"
v-loading="tableLoading" :element-tableLoading-text="$t('action.loading')">
            <el-table-column prop="title" :label="$t('common.versionName')" header-align="center" align="center">
            </el-table-column>
            <el-table-column fixed="right" :label="$t('action.operation')" width="180">
                <template slot-scope="scope">
                    <el-button @click="handleRestore(scope.row)" type="primary" size="mini">{{$t('common.restore')}}</el-button>
                    <el-button @click="handleDelete(scope.row)" type="danger" :disabled="scope.row.name=='backup'?true:false" size="mini">{{$t('action.delete')}}</el-button>
                </template>
```

```
        </el-table-column>
      </el-table>
      <span slot="footer" class="dialog-footer">
        <el-button size="small" @click="backupVisible = false">{{$t('action.cancel')}}</el-button>
        <el-button size="small" type="primary" @click="handleBackup">{{$t('common.backup')}}</el-button>
      </span>
    </el-dialog>
</template>
```

针对备份查询、备份创建、备份还原和备份删除的方法如下，因为备份还原服务比较简单，这里就没有对 axios 进行封装了，直接调用 axios 方法即可。

### 2. 备份查询

```
views/Backup/Backup.vue
findRecords: function () {
    this.tableLoading = true
    axios.get(this.baseUrl + '/backup/findRecords').then((res) => {
        res = res.data
        if(res.code == 200) {
            this.tableData = res.data
        } else {
            this.$message({message: '操作失败, ' + res.msg, type: 'error'})
        }
        this.tableLoading = false
    })
}
```

### 3. 备份创建

```
views/Backup/Backup.vue
handleBackup: function () {
    this.tableLoading = true
    axios.get(this.baseUrl + '/backup/backup').then((res) => {
        res = res.data
        if(res.code == 200) {
            this.$message({ message: '操作成功', type: 'success' })
        } else {
            this.$message({message: '操作失败, ' + res.msg, type: 'error'})
        }
        this.tableLoading = false
        this.findRecords()
    })
}
```

### 4. 备份还原

```
views/Backup/Backup.vue
handleRestore: function (data) {
    this.tableLoading = true
    axios.get(this.baseUrl + '/backup/restore', {params : {name : data.name }}).then((res) => {
        res = res.data
        if(res.code == 200) {
            this.$message({ message: '操作成功', type: 'success' })
            this.$emit('afterRestore', {})
        } else {
            this.$message({message: '操作失败, ' + res.msg, type: 'error'})
        }
        this.tableLoading = false
    })
}
```

### 5. 备份删除

```
views/Backup/Backup.vue
handleDelete: function (data) {
    this.tableLoading = true
    axios.get(this.baseUrl + '/backup/delete', {params : {name : data.name }}).then((res) => {
        res = res.data
        if(res.code == 200) {
            this.$message({ message: '操作成功', type: 'success' })
        } else {
            this.$message({message: '操作失败, ' + res.msg, type: 'error'})
        }
        this.findRecords()
        this.tableLoading = false
    })
}
```

## 32.4 页面引用

我们之前在用户面板预留了备份还原的操作入口，在用户面板中引入备份还原组件。views/Core/PersonalPanel.vue 相关内容如图 32-2 所示。

```
</el-dialog>
<!--备份还原界面-->
<backup ref="backupDialog" @afterRestore="afterRestore"></backup>
</div>
</template>

<script>
import Backup from "@/views/Backup/Backup"
import { format } from "@/utils/datetime"
export default {
  name: 'PersonalPanel',
  components:{
    Backup
  },
```

图 32-2

在备份还原操作中添加响应函数，打开备份还原界面。views/Core/PersonalPanel.vue 相关内容如图 32-3 所示。

```
<div class="other-operation-item" @click="openOnlinePage">
  <li class="fa fa-user"></li>
  在线人数 {{onlineUser}}
</div>
<div class="other-operation-item">
  <li class="fa fa-bell"></li>
  访问次数 {{accessTimes}}
</div>
<div class="other-operation-item" @click="showBackupDialog">
  <li class="fa fa-undo"></li>
  {{$t("common.backupRestore")}}
</div>
```

图 32-3

响应函数，打开备份还原界面：

```
views/Core/PersonalPanel.vue
// 打开备份还原界面
showBackupDialog: function() {
  this.$refs.backupDialog.setBackupVisible(true)
}
```

还原操作响应函数，还原成功之后清除用户信息并返回到登录页面。

```
views/Core/PersonalPanel.vue
// 成功还原之后，重新登录
afterRestore: function() {
  this.$refs.backupDialog.setBackupVisible(false)
  sessionStorage.removeItem("user")
  this.$router.push("/login")
  this.$api.login.logout().then((res) => {
  }).catch(function(res) {
  })
}
```

## 32.5 页面测试

启动注册中心、监控服务、后台服务、备份服务，访问 http://localhost:8080，登录应用进入主页，单击"用户面板"→"备份还原"，弹出备份还原界面。

#### 1. 备份查询

可以看到初始状态系统会为我们创建一个初始备份，即系统默认备份，且系统默认备份不可删除，防止所有备份被删除时无备份可用，如图 32-4 所示。

图 32-4

#### 2. 备份创建

单击"备份"按钮，进行数据备份，如图 32-5 所示，系统创建了一个新的备份。备份名称格式为"backup_ + 时间戳"，对应 SQL 文件存储目录名称。

#### 3. 备份删除

单击"删除"按钮，如图 32-6 所示，备份被成功删除。

图 32-5

图 32-6

### 4. 备份还原

接下来我们测试一下数据还原功能，选择"系统管理"→"机构管理"，删除轻尘集团和牧尘集团的两棵机构树，只剩下三国集团，如图 32-7 所示。

单击"用户面板"→"备份还原"→"还原"按钮，进行数据还原，如图 32-8 所示。

图 32-7

图 32-8

还原成功之后，会回到登录页面，重新登录，查看机构管理，发现数据已经恢复回来了，如图 32-9 所示。

图 32-9